Biomedical Applications of Extracellular Vesicles

Biomedical Applications of Extracellular Vesicles

Edited by Zhenhua Li, Xing-Jie Liang, and Ke Cheng

WILEY-VCH

Editors

Prof. Zhenhua Li
The Tenth Affiliated Hospital of
Southern Medical University
Dongguan
China

Prof. Xing-Jie Liang
National Center for Nanoscience and
Technology
Beijing
China

Prof. Ke Cheng
Columbia University
New York
NY, US

Cover: © Jason Drees/Shutterstock;
Marcus Millo/Getty Images/iStockphoto

Library of Congress Card No.: applied for

British Library Cataloguing-in-Publication Data
A catalogue record for this book is available from the British Library.

Bibliographic information published by the Deutsche Nationalbibliothek
The Deutsche Nationalbibliothek lists this publication in the Deutsche Nationalbibliografie; detailed bibliographic data are available on the Internet at <http://dnb.d-nb.de>.

Print ISBN: 978-3-527-35212-8
ePDF ISBN: 978-3-527-84213-1
ePub ISBN: 978-3-527-84214-8
oBook ISBN: 978-3-527-84215-5

Typesetting: Straive, Chennai, India
Printing and Binding: CPI Group (UK) Ltd, Croydon, CR0 4YY

C9783527352128_241123

Contents

Preface

Extracellular vesicles (EVs) are a type of nano-sized particles that are secreted by cells and widely found in tissues, serum, and body fluids. In the past decades, EVs have drawn increasing interest because of their involvement in various physiological and pathological activities, such as cell–cell communication, cell death, and immune regulation. The understanding of EV features and biofunctions also facilitates the progress of biomedical applications of EVs. For instance, EVs have been used as drug delivery systems due to their inherent ability of substance transportation. In addition, scientists have conducted various attempts aiming to achieve the clinical and industrial translation of EV-based therapeutics. The purpose of this book is to provide a detailed image of the present understanding of EVs and derived artificial nanovesicles and their biomedical applications.

This book includes eight chapters that focus on different aspects of EVs. Each chapter introduces the current knowledge of EVs in a specific area of interest and discusses the possible future direction. In Chapter 1, the authors provide a brief introduction to EVs, including their nature and biological functions. In Chapter 2, the authors focus on EV biogenesis and identification. Specifically, the biogenesis of various EVs, including exosomes, microvesicles, apoptotic bodies, and large oncosomes, are systematically summarized. In addition, current techniques in the identification of EVs, such as electron microscopy, nanoparticle tracking analysis, and dynamic light scattering are introduced. Chapter 3 mainly discusses the biological functions of EVs from different sources, such as stem cells, immune cells, tumor cells, plants, and microorganisms. Their applications in anti-inflammation, tissue regeneration, and neuroprotection are also discussed. In Chapter 4, the authors summarize the recent achievements in the applications of EVs in disease therapy. Several major diseases, including cardiovascular diseases, tissue engineering and regenerative medicine, cancer, respiratory diseases, and metabolic diseases, are selected to reveal their therapeutic efficacy. To further enhance the therapeutic efficiency and reduce the side effects of EVs, researchers have also been devoted to developing novel engineered EVs via various techniques. The engineered EVs for biomedical applications are discussed in Chapter 5. In Chapter 6, the authors introduce the current approaches for the production, isolation, and quality control of EVs. Chapter 7 summarizes the current progress of EVs in clinical applications. In addition, the potential challenges and future trends of clinical translations of EV are discussed.

In Chapter 8, a summary of the main contents of this book is provided as a concluding chapter.

Taken together, this book summarizes the current knowledge and potential applications of EVs across various basic and technical disciplines. We further discussed the challenges in the current EV research, and their translation in clinical and industrial research. We hope this book could act as a handbook to help scientists in the field of EV study and promote the development of EV research. We would like to give our sincere gratitude to all the authors, the reviewers, and the editors who have contributed and assisted in this book. Without their valuable time and efforts, this book would never be possible.

Dongguan, Guangdong, China, Zhenhua Li
Beijing, China, Xing-Jie Liang
New York, NY, USA Ke Cheng

1

Extracellular Vesicles and Their Biomedical Applications: An Overview

Xing-Jie Liang[1], *Ke Cheng*[2], *and Zhenhua Li*[3,4]

[1] *CAS Key Laboratory for Biological Effects of Nanomaterials and Nanosafety, National Center for Nanoscience and Technology, Beijing, 100190, China*
[2] *Department of Biomedical Engineering, Columbia University, New York, 10032, USA*
[3] *The Tenth Affiliated Hospital of Southern Medical University, 78 Wanjiang Avenue, Dongguan, Guangdong 523059, China*
[4] *Guangdong Provincial Key Laboratory of Shock and Microcirculation, Southern Medical University, Shatai South Avenue Guangzhou, Guangdong 510080, China*

1.1 Introduction

Extracellular vesicles (EVs) are small membrane-bound vesicles that are secreted by a variety of cells, including stem cells, immune cells, and cancer cells. These vesicles contain a range of molecules, including lipids, proteins, and nucleic acids, that reflect the cellular and physiological state of the parent cell, and play a variety of physiological and pathological roles in the body. They can act as signaling molecules, carrying bioactive molecules such as proteins, lipids, and nucleic acids, between cells and tissues. They can also act as vehicles for the transfer of genetic material, such as microRNAs, between cells. The biomedical applications of EVs have been widely explored in recent years, and they are emerging as a promising tool for diagnosis, therapy, and drug delivery.

1.2 Biogenesis and Composition of Extracellular Vesicles

EVs are released by cells into the extracellular space and can be classified into three main types based on their biogenesis and size: exosomes, microvesicles, and apoptotic bodies. Exosomes are small vesicles (30–150 nm) that are released by cells through the endosomal pathway, while microvesicles (100–1000 nm) are formed by budding of the plasma membrane. Exosomes and microvesicles are collectively referred to as small EVs. Exosomes are enriched in endosomal markers such as CD63 and Alix, while microvesicles are enriched in plasma membrane markers such as phosphatidylserine and CD31. Apoptotic bodies are referred to as large EVs (1–5 μm)

Biomedical Applications of Extracellular Vesicles, First Edition.
Edited by Zhenhua Li, Xing-Jie Liang, and Ke Cheng.

that are released by cells undergoing programmed cell death. The biogenesis of each type of EV is distinct and involves the selective packaging of different molecules.

The composition of EVs varies depending on the type of vesicle, the cell of origin, and the physiological or pathological condition. EVs can contain a range of biological molecules, including proteins, nucleic acids (RNA and DNA), lipids, and carbohydrates. Proteins present in EVs include membrane proteins, cytosolic proteins, and extracellular matrix proteins. Nucleic acids in EVs include mRNAs, miRNAs, and long noncoding RNAs. Lipids in EVs include phospholipids, sphingolipids, and cholesterol.

1.3 Biological Functions of Extracellular Vesicles

EVs have been shown to regulate a range of cellular processes, including proliferation, differentiation, and apoptosis. EVs have been shown to play a role in many biological processes, including immune regulation, inflammation, angiogenesis, tissue repair, and cancer progression. They have also been implicated in the pathogenesis of many diseases, including cardiovascular disease, neurological disorders, and infectious diseases. Given their diverse functions, EVs have the potential to be used for a range of biomedical applications, especially in the fields of diagnostics, therapeutics, and regenerative medicine.

In the area of diagnostics, EVs have great potential as diagnostic biomarkers for a variety of diseases. They are stable in biological fluids, and their cargo can reflect the physiological state of the cell of origin, making them attractive targets for disease diagnosis and monitoring. Cancer is one area where EVs have shown particular promise as diagnostic biomarkers. Cancer cells release large numbers of EVs into the circulation, which can be detected in blood samples. These EVs contain specific biomolecules that can be used to detect and monitor the progression of cancer. For example, circulating tumor cells (CTCs) shed EVs that contain proteins and nucleic acids that are specific to the cancer cells. These biomolecules can be used to develop noninvasive diagnostic tests for cancer. In addition, EVs released by cancer cells can provide information about the molecular profile of the tumor, which can be used to develop personalized cancer treatments. EVs have also been explored as diagnostic biomarkers for other diseases, such as cardiovascular disease, neurological disorders, and infectious diseases. For example, EVs released by damaged or diseased heart tissue contain specific proteins and nucleic acids that can be used to diagnose and monitor cardiovascular disease.

Recent research has shown that EVs have enormous potential as therapeutic agents for a variety of diseases, including cancers, neurodegenerative diseases, and cardiovascular disease. Here are some of the key ways that EVs are being investigated for therapeutic use. In cancer therapy, EVs derived from stem cells have been shown to inhibit tumor growth and metastasis in several animal models of cancer. EVs have also been explored as a means of drug delivery, as they can be engineered to carry therapeutic molecules directly to tumor cells. In neurodegenerative diseases, EVs derived from stem cells have shown promise in treating neurodegenerative diseases such as Parkinson's and Alzheimer's. These EVs contain factors that can protect neurons from damage and promote their survival. In cardiovascular disease, EVs derived from endothelial cells have been shown to improve heart function after a heart attack in animal models. EVs can also carry proteins and genetic materials that promote angiogenesis, which can help to repair damaged blood vessels. In autoimmune diseases, EVs have been shown to play a role in regulating immune responses, and EVs derived from stem cells have been explored as a potential treatment for autoimmune diseases such as multiple sclerosis and rheumatoid arthritis. Overall, EVs offer a promising avenue for the development of novel therapeutics for a wide range of diseases. While there are still many challenges to overcome, including standardization of isolation and characterization techniques, early results suggest that EVs may be a powerful tool in the fight against disease.

EVs have been shown to have tremendous potential in regenerative medicine. EVs are small, membrane-bound structures secreted by cells that contain a variety of biomolecules, including lipids, proteins, and nucleic acids. They play a critical role in intercellular communication and have been shown to have therapeutic effects in a variety of preclinical models of disease. Here are some of the key ways that EVs are being investigated for regenerative medicine. In tissue repair, EVs derived from stem cells have been shown to promote tissue repair in a variety of preclinical models, including wound healing, bone regeneration, and cartilage repair. These EVs contain a variety of growth factors and other molecules that can promote cell proliferation and differentiation, as well as regulate the immune response. In cardiovascular regeneration, EVs have been shown to have beneficial effects on cardiac function following a heart attack in animal models. EVs can carry proteins and genetic material that promote angiogenesis, or the growth of new blood vessels, which can help repair damaged tissue and improve blood flow. In neuroregeneration, EVs have shown promise in promoting neuronal survival and regeneration in preclinical models of neurological disease and injury, including stroke and traumatic brain injury. These EVs can carry neuroprotective factors, and promote the growth and differentiation of new neurons. Overall, EVs offer a promising approach to regenerative medicine, while there are still challenges to overcome, including standardization of isolation and characterization techniques.

1.4 Extracellular Vesicles Isolation and Limitations

EVs have attracted considerable attention as potential biomarkers for various diseases, including cancer, cardiovascular disease, and neurodegenerative disorders. However, the isolation of EVs is a complex and challenging process, and there are several limitations in the manufacturing and clinical translation of EV-based therapeutics.

Isolation of EVs is a crucial step in their study and application. There are several methods for isolating EVs, including ultracentrifugation, size-exclusion chromatography, immunoaffinity capture, and microfluidics-based techniques. Each method has its advantages and limitations in terms of purity, yield, and scalability. Ultracentrifugation is the most commonly used method for EV isolation, but it can result in coisolation of non-EV contaminants, such as protein aggregates and lipoproteins. Size-exclusion chromatography can provide higher purity, but it is less efficient and requires specialized equipment. Immunoaffinity capture can specifically isolate EVs expressing certain surface markers, but it can result in low yields and requires specific antibodies. Microfluidics-based techniques have the potential for high throughput and precise isolation, but they are still in the development stage and not yet widely used.

There are also some limitations in both manufacturing and clinical translation. One of the major limitations in manufacturing EV-based therapeutics is the variability in EV isolation methods that can lead to variations in the size, content, and purity of EVs. This can impact the efficacy and safety of EV-based therapeutics. Moreover, the scalability of EV isolation methods is also a challenge, as large-scale production of EVs is currently not feasible. Another challenge in manufacturing EV-based therapeutics is the lack of standardized protocols for EV characterization, which can lead to inconsistencies in the reporting of EV-related data. This can make it difficult to compare data between different studies and hinder the development of EV-based therapeutics. As for the clinical translation of EV-based therapeutics is the lack of a regulatory framework for EVs, there is currently no FDA-approved EV-based therapy, and the regulatory pathway for EVs is not well defined. This can make it challenging for researchers and companies to develop and commercialize EV-based therapeutics. Another challenge in clinical translation is the heterogeneity of EVs, which can make it difficult to define a specific population of EVs for therapeutic use. Moreover, the biodistribution and pharmacokinetics of EVs are not well understood, and there is limited data on the safety and efficacy of EV-based therapeutics in humans.

In conclusion, while EVs hold great potential as biomarkers and therapeutics for various diseases, there are several challenges in the isolation, manufacturing, and clinical translation of EV-based products. These challenges highlight the need for standardized protocols, regulatory guidance, and further research to fully harness the potential of EVs for clinical use.

2

Biogenesis and Identification of Extracellular Vesicles

Dandan Ding[1,2], Xing Zhang[3], Yu Zhao[3], Xiaoya Li[3], Qingqing Leng[3], and Zhenhua Li[1,2]

[1] The Tenth Affiliated Hospital of Southern Medical University, 78 Wanjiang Avenue, Dongguan, Guangdong 523059, China
[2] Guangdong Provincial Key Laboratory of Shock and Microcirculation, Shatai South Avenue, Guangzhou, Guangdong 510080, China
[3] Hebei University, College of Pharmaceutical Science, Key Laboratory of Pharmaceutical Quality Control of Hebei Province and Key Laboratory of Medicinal Chemistry and Molecular Diagnosis of Ministry of Education, Chemical Biology Key Laboratory of Hebei Province, Institute of Life Science and Green Development, Baoding 071002, China

Extracellular vesicles (EVs) play an important role in physiological and pathological processes as widely available bioactive substances involved in intercellular information exchange [1–3]. Compared with conventional nanomaterials, EVs are one of the most promising candidates in nanomedicine because of their biocompatibility, biodegradability, low toxicity, and non-immunogenicity. As an emerging research field in recent years, it has been found that EVs can be involved in the occurrence and development of a variety of diseases and have great potential as markers of diseases [4, 5]. With in-depth research, the exploration of their biogenesis is important for the early diagnosis and treatment of diseases. In addition, since the existing isolation techniques are based on the size and structure of EVs, and the capture of some membrane proteins, it is difficult to completely distinguish them from other vesicles and macromolecular protein complexes, which requires their identification. Chapter 2 focuses on EVs through biogenesis mechanism and identification.

2.1 Biogenesis of Extracellular Vesicles

On the basis of their biogenesis, cargo, and size, EVs are generally divided into four broad categories: exosomes, microvesicles (MVs), apoptotic bodies (ApoBDs), and large oncosomes (LOs). Exosomes (30–200 nm in size) and MVs (200–2000 nm in size) are produced by almost all kinds of healthy living cells, derived from the fusion of multivesicular bodies (MVBs) with the plasma membrane, and formed by a direct outward budding of the plasma membrane, respectively [6]. ApoBDs

Biomedical Applications of Extracellular Vesicles, First Edition.
Edited by Zhenhua Li, Xing-Jie Liang, and Ke Cheng.

(500–2000 nm in size) are released by apoptotic cells during programmed cell death [7]. LOs (1–10 μm in size) are cancer cell-derived microvesicles [8].

2.1.1 Biogenesis of Exosome

Exosomes are usually generated through the endosome pathway [9]. First, the invagination of the plasma membrane leads to the formation of endosomes. Then, intraluminal vesicles (ILVs) were further formed by the inward budding of the limiting membrane inside endosomes. ILVs further accumulate in the lumen to form multivesicular bodies (MVBs) [1, 10]. Eventually, MVBs may be degraded by fusion with lysosomes [11]. Alternatively, MVB fuses with the plasma membrane and releases ILVs into the extracellular environment in the form of exosomes [12].

The transformation of endosomes into MVBs containing ILVs is an important step of exosome biogenesis. The main mechanism of ILVs' formation is the endosomal sorting complex required for transports (ESCRTs) [13]. ESCRTs include four different protein complexes (ESCRT-0, ESCRT-I, ESCRT-II, and ESCRT-III) [14–17]. In the process of ILVs formation, ESCRT complexes are gradually recruited to the endosomal membrane. First, ESCRT-0 is recruited to the early endosomal membrane via direct interaction of the hepatocyte growth factor-regulated tyrosine kinase substrate (HRS) with phosphatidylinositol 3-phosphate (PIP3) on the endosomal membrane [18–21]. HRS can recognize and internalize the ubiquitination proteins [22, 23]. Then, ESCRT-0 recruits ESCRT-I to endosomal membranes by HRS binding to tumor susceptibility gene 101 (*TSG*101) of ESCRT-I [24, 25]. ESCRT-0 and ESCRT-I are jointly responsible for the sorting of ubiquitinated proteins [26]. ESCRT-II binds to ubiquitin, and ESCRT-I binds to Vps36 subunit [27–29]. ESCRT-I/ESCRT-II complex has been thought to play a role in the inward budding of the endosomal membrane [27, 30, 31]. ESCRT-II activates ESCRT-III to bind to the budding neck [32, 33]. Finally, AAA-ATPase Vps4 drives the plasma membrane separation, generating ILVs, and leading to the formation of MVBs [34–36].

Several other ESCRT mechanisms have been demonstrated in the formation of ILVs. (i) The syndecan–syntenin–ALIX exosome biogenesis pathway has been proposed for ILVs formation without the involvement of ubiquitination and ESCRT-0, but dependent on ESCRT-III [37, 38]. In this pathway, biogenesis of the exosome and loading of proteins are regulated by heparanase, syndecan heparan sulfate proteoglycans, the small GTPase ADP ribosylation factor 6 (ARF6), and phospholipase D2 (PLD2) [38–40]. (ii) His domain protein tyrosine phosphatase (HD-PTP) acts as a scaffold to continuously recruit ESCRT-0, ESCRT-I, and ESCRT-III without the involvement of ESCRT-II [41–43]. (iii) ALIX can directly bind protease-activated receptor 1 (PAR1) or lysobisphosphatidic acid (LBPA) to recruit ESCRT-III independent of ESCRT-0, ESCRT-I, and ESCRT-II [44–47].

In addition, it has been shown that the existence of mechanisms for exosome biogenesis is not dependent on ESCRT. For example, the formation of MVB cannot be inhibited in the absence of the ESCRT complex [48]. Trajkovic et al. find that the formation of ILVs is independent of ESCRT but requires the sphingolipid ceramide [49]. The cone-shaped structure of ceramide promotes membrane invagination of the endosomal membrane [50]. The release of exosomes is decreased after inhibition of neutral sphingomyelinase (nSMase), which can produce ceramides [51–53]. Another ESCRT-independent exosome biogenesis is mediated by tetraspanins. In this mechanism, exosome secretion is not inhibited even in the absence of ESCRTs and ceramides [54]. In a tetraspanin-dependent mechanism, tetraspanins can aggregate related molecules to endosomal membranes, leading to the invagination of the endosomal membrane to form ILVs [55, 56]. Tetraspanin CD63 plays an important role in exosome formation and release [57–59]. In addition, CD9, CD81, and CD82 have also been shown to be involved in the formation of exosomes [60, 61].

After the formation of ILVs, MVBs can be degraded by fusing with lysosomes. In this mechanism, ISGylation of ESCRT-I component TSG101 can reduce exosome secretion by promoting the fusion of MVBs with lysosomes [62, 63]. Another fate of MVBs is promoting exosome release into the extracellular environment by fusing with the plasma membrane. The RAB GTPases, including RAB7, RAB11, RAB27, RAB31, and RAB35, are thought to play a crucial role in the translocation of MVBs to the plasma membrane [64, 65]. It was shown that deletion of RAB7 reduced the secretion of exosomes containing syntenin and ALIX in MCF-7 cells, but did not affect the release of exosomes from Hela cells [37, 66]. Recently, Fei et al. demonstrated that RAB7 could be recruited to MVBs by binding directly to neddylated Coro1a, promoting degradation of MVBs and leading to a decrease in exosome secretion [67]. RAB11 promotes the release of exosomes containing transferrin receptor (TfR) and heat shock cognate 71 kDa protein (Hsc70) in K562 cells [68]. Both subtypes of RAB27— RAB27A and RAB27B—affect exosome release from cancer cells [69–71]. Silencing of RAB27A leads to decreased secretion of exosomes containing the traditional markers CD63, TSG101, ALIX, and HSC70, without affecting the secretion of exosomes containing CD9 and Mfge8 [72]. RAB31—an ESCRT-independent exosome biogenesis pathway—drives the formation of ILVs and prevents the fusion of MVBs with lysosomes [73]. RAB35 regulates the release of exosomes from oligodendrocytes [74]. Inhibition of RAB35 reduced the release of proteolipid protein (PLP)-containing exosomes from oligodendrocytes [74].

The final step in exosome secretion is controlled by the soluble NSF attachment protein receptor (SNARE) protein. SNARE protein can mediate membrane fusion of different intracellular compartments [75–79]. Vesicle-associated membrane protein 7 (VAMP7) is involved in Ca^{2+}-regulated exocytosis of traditional lysosomes and regulates exosome release from K562 cells [80–82]. Syntaxin 1A (Syx1A) is a necessary SNARE protein for evenness interrupted (Evi)-containing exosome secretion [83]. In non-small cell lung cancer, the SNARE protein YKT6 is essential for regulating Wnt-containing exosome secretion [84, 85]. The target membrane SNARE protein SYX-5 also participates in regulating exosome secretion; MVBs aggregate under the plasma membrane when SYX-5 is deficient [86]. In addition, SNAP-23 promotes exosome secretion in tumor cells by regulating pyruvate kinase type M2 (PKM2)-mediated phosphorylation [87, 88].

2.1.2 Biogenesis of Microvesicle

Microvesicles (MVs) are generated mainly by direct outward budding and division of the plasma membrane, and have multiple biogenesis and release mechanisms. The redistribution of phospholipids and contraction of the cytoskeleton through actin–myosin interactions together lead to the formation of MVs [6, 89–92]. Recent studies have shown that ESCRT is involved in the biogenesis of MVs [93, 94]. The arrestin domain-containing protein 1 (ARRDC1) facilitates the production of MVs containing TSG101, ARRDC1, and intracellular proteins by binding to ESCRT-I subunit TSG101 and recruiting it to the plasma membrane [95]. Vps4 ATPase is also required for the generation of such MVs [95]. Activation of acid sphingomyelinase induces ceramide-dependent production of MVs from astrocytes [96]. Similarly, acidic sphingomyelinase also enhances the production of MVs in erythrocytes [97]. The formation of MVs is also correlated with cholesterol levels, and the production of MVs significantly decreases when cholesterol is depleted [98]. Caveolin-1 (cav-1) has been reported to be involved in the formation of miRNA-containing MVs [99]. Phosphorylation of cav-1 tyrosine 14 (Y14) leads to the interaction of cav-1 with hnRNPA2B1, which induces hnRNPA2B1 to bind miRNAs, resulting in selection of miRNAs into MVs [99]. Other than the above mechanism, small GTPase proteins, such as PhoA, ARF1, and ARF6, can also regulate production of MVs [92, 100, 101]. In addition, external stimuli can also induce the release of MVs. In erythrocytes and platelet cells, calcium influx induces phospholipid redistribution and promotes the release of MVs [102, 103]. Hypoxia also facilitates the release of MVs from breast cancer by the HIF-dependent expression of PAB22A [104].

2.1.3 Biogenesis of Apoptotic Bodies

Apoptotic bodies (ApoBDs) are generated only during programmed cell death [6]. ApoBDs originate from the condensation of nuclear chromatin, undergo nuclear division and membrane blebbing, and eventually the cell contents split into distinct membrane-encapsulated vesicles referred to as ApoBDs [105, 106]. At present, applied studies on ApoBDs are scarce, mainly attributed to the limited understanding of the biogenesis of ApoBDs. Membrane blebbing is a key morphological step in the formation of ApoBDs. Studies suggest that membrane blebbing may be mediated by actin–myosin interactions and is thought to be regulated by a number of kinases, including Rho-associated kinase 1 (ROCK1) [107–109], p21 activated kinase 2 (PAK2) [110, 111], and Lim domain kinase 1(LIMK1) [112]. In addition, in T cells and thymocytes, the formation of ApoBDs is regulated by the cystatin-activated pannexin 1 (PANX1) channel. Blocking the activity of PANX1 channel promotes the formation of ApoBDs [113]. ApoBDs can promote phagocytosis and clearance of apoptotic cells by macrophages [114–116]. During apoptosis, membrane lipids undergo rearrangement, transferring phosphatidylserine located in the inner leaflet of the plasma membrane to the outer leaflet [117], facilitating the binding of phosphatidylserine to Annexin V of macrophages [118], and achieving the clearance of apoptotic cells.

2.1.4 Biogenesis of Large Oncosomes

Large oncosomes (LOs) are secreted by cancer cells in an amoeboid motile manner. Di Vizio et al. revealed that amoeboid migration of DU145 and LNCaP human prostate cancer cells triggered the generation of giant EVs referred to as Los [119]. It was shown that extracellular matrix degradation products, such as elastin degradation products (EDPs), affect intracellular calcium flow and cytoskeleton reorganization, inducing a tumor amoeboid phenotype [120]. LOs are byproducts of nonapoptotic cell plasma membrane blebbing, and their shedding process is regulated by various proteins [121]. The shedding of LOs can be promoted by silencing of the cytoskeletal regulator diaphanous-related formin-3 (DIAPH3) protein, overexpression of oncoproteins including caveolin-1 (Cav-1), myristylated Akt1 (MyrAkt1) and heparin-binding epidermal growth factor (HB-EGF), or activation of EGFR [8, 119, 122–124]. In addition, cytokeratin 18 (CK18) can be used as a marker of tumor-derived LOs [124].

2.2 Identification of Extracellular Vesicles

2.2.1 Electron Microscopic Identification

Electron microscopy has a high resolution and can directly observe the morphology and structure of nanoparticles in a sample, which can therefore be used to identify the presence and integrity of EVs; but the electron micrographs of nanoparticles such as lysosomes, mycoplasma, polyethylene glycol protein aggregates, and ferritin aggregates are very similar to those of exosomes, and the electron microscope can only show a partial view. Therefore, it cannot distinguish between EVs and nanoparticles with similar morphology, nor does it reflect the number of EVs in the sample. Moreover, electron microscopy instruments are expensive, the method requires more stringent sample preparation, and the results are subject to subjectivity [125].

2.2.1.1 Scanning Electron Microscopy

Scanning electron microscope (SEM) is the use of secondary electron signal imaging to observe the surface morphology of a sample. SEM is a microscopic morphology observation tool between transmission electron microscopy and optical microscopy, which can directly image through the surface properties of the material. The images obtained by SEM are stereoscopic and can be used to observe various morphological features of biological samples, mainly for observing the surface morphology of nanomaterials, analysis of material fractures, direct observation of the original surface, observation of thick specimens, and observation of the details of each area. The resolution of advanced SEM is less than 1 nm, which can meet the requirements of EV size observation. The EVs observed using SEM show a distorted cup-like morphology, which may be due to the deformation of the vesicles as a result of chemical fixation and dehydration treatment required for sample preparation.

As we can see from Figure 2.1A, Chukhchin et al. studied EVs from the phloem and xylem of woody plants by SEM, which revealed the cup-like structural features

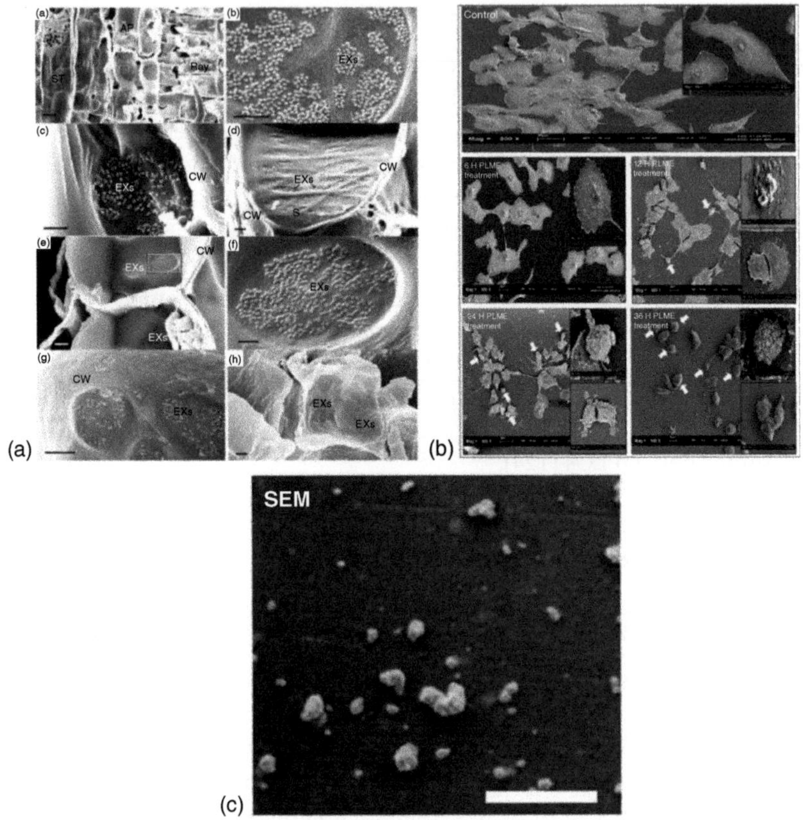

Figure 2.1 (a) Exosomes in the secondary phloem of woody plants. Source: Chukhchin et al. [126]/with permission of Springer Nature. (b) SEM micrographs of surface ultrastructure characteristics of HeLa cells. Source: Vijayarathna et al. [127]/with permission of Elsevier. (c) SEM image of a selected ROI. Source: Beekman et al. [128]/with permission of Royal Society of Chemistry.

of EVs in the dry state [126]. Vijayarathna et al. observed the generation and morphological features of apoptotic vesicles by scanning electron microscopy (Figure 2.1B) [127]. Beekman et al. analyzed EVs of tumor origin using scanning electron microscopy and studied the size and size distribution of individual EVs levels (Figure 2.1C) [128].

2.2.1.2 Transmission Electron Microscopy

Transmission electron microscope (TEM) is the most widely used type of electron microscope. TEM is used to observe and study the internal submicroscopic structure of cells and the morphological structure of viruses, proteins, nucleic acids, and other biological macromolecules. [129]

TEM is an electro-optical instrument with high resolution and high magnification that can be used to determine whether the isolated and purified sample has a more typical structure of EVs, distinguishing EVs from nanoparticles. Currently, the

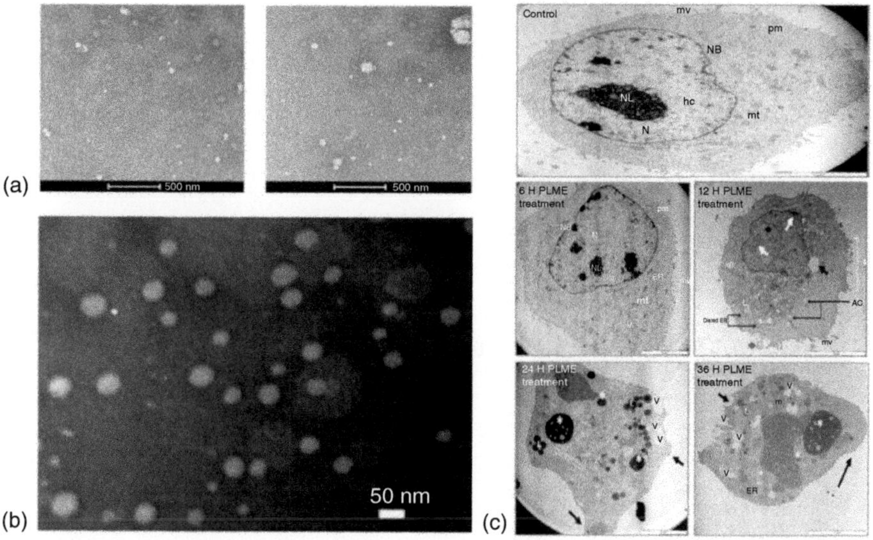

Figure 2.2 TEM picture of EVs. Source: (a) Wang et al. [130]/Journal of Immunology Research/CC BY 4.0.; (b) Yuan et al. [131]/with permission of Taylor and Francis group.; (c) Vijayarathna et al. [127]/with permission of Elsevier.

resolution of the most advanced TEM has reached 0.1 nm. However, the number of particles observed at one time is limited and is not suitable for large sample sizes. EVs observed by TEM appear as cups with a double-layer membrane structure. Similar to SEM, the process of fixation and dehydration of the sample prior to observation leads to unpredictable changes in the structure of EVs.

As we can see from Figure 2.2a, Wang et al. verified the morphological and structural characteristics of EVs from PC cells, and their internalization by human umbilical vein endothelial cells (HUVEC) by TEM [130]. Yuan et al. prepared a self-assembled vesicle in aqueous solution, and confirmed that the vesicles were spherical in shape using TEM and had an average particle size of ~33.7 nm as measured by dynamic light scattering (Figure 2.2b) [131]. Vijayarathna et al. used TEM to confirm the extrusion of cytoplasm and the formation of apoptotic vesicles (Figure 2.2c) [127].

2.2.1.3 Atomic Force Microscopy

Atomic force microscopy (AFM) can be used to study the surface structure of solid materials, including insulators. It studies the surface structure and properties of materials by detecting extremely weak interatomic interaction forces between the surface of the sample to be measured and a micro-force-sensitive element [132].

Compared to other electron microscopes, AFM has been found to have some significant advantages in the identification of EVs. Firstly, the operating condition of AFM is not harsh; it can work at atmospheric pressure and even in liquid environments. This allows it to be used to study macroscopic characteristics of organisms and even living biological tissues. Secondly, AFM can form a true

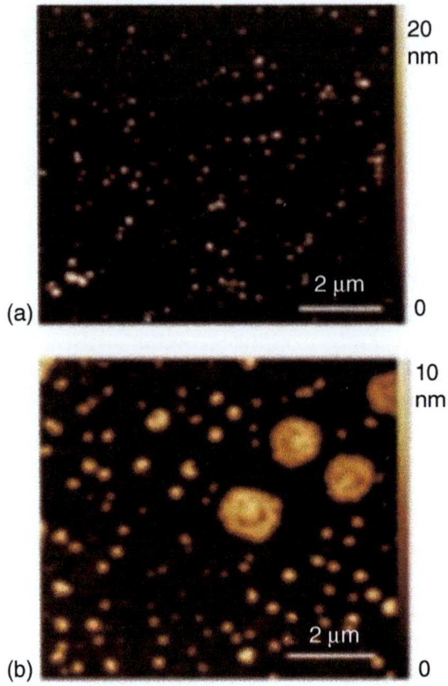

Figure 2.3 EV surface structure observed by AFM. Source: (a) Sharma et al. [133]/with permission of American Chemical Society.; (b) Ali et al. [134]/with permission of Elsevier.; (c) Sharma et al. [135]/Springer Nature/CC BY 4.0.

three-dimensional structure, which is more convenient for observation. Finally, AFM can be observed without destructive processing of the sample, which ensures the accuracy of the observation results. Compared with SEM, the disadvantages of AFM are that the imaging range is too small, the speed is slower, and the influence of the probe is too great.

As Figure 2.3A shows, Sharma et al. used AFM to demonstrate the structure of EVs that could not be clearly identified under electron microscopy and correlated the data with field emission scanning electron microscopy (FESEM) and AFM images to explain the nanoscale structure of EVs under different forces [133]. Single EVs showed reversible mechanical deformation, displaying distinctive 70–100 nm elastic trilobal membranes with substructures bearing specific transmembrane receptors (Figure 2.3B). Ayat et al. used AFM to detect changes in cell surface morphology, granulation, and mean surface roughness, and found cell shrinkage and increased cytoplasmic organelles, confirming the generation of apoptotic vesicles [134]. Sharma et al. compared the size, structure, and surface properties of small EVs (sEVs) derived from breast cancer cells using AFM (Figure 2.3C) [135].

2.2.1.4 Cryo-Electron Microscopy

Cryo-electron microscopy (Cryo-M) is an ultra-low temperature frozen sample preparation and delivery technique for scanning electron microscopy, which could provide direct observation of liquid, semiliquid, and beam-sensitive samples, such as biological and polymeric materials. Cryo-M can analyze the morphology of EVs without extensive processing. When using Cryo-M for observation of samples, the vacuum environment and the impact of electron energy cause rapid deterioration

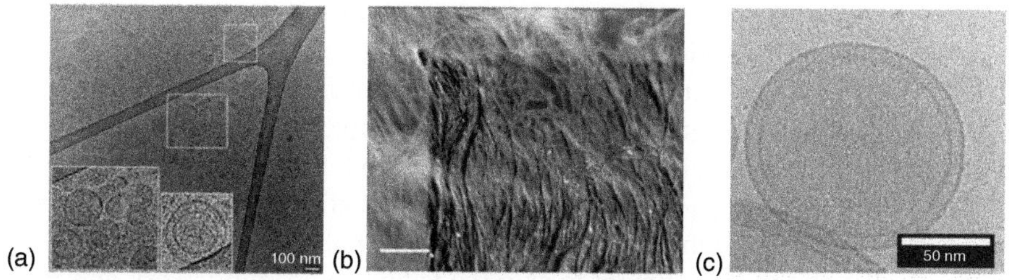

Figure 2.4 Picture of inflammatory EVs. Source: (a) Bairamukov et al. [137]/with permission of Elsevier.; (b) Matatyaho Ya'akobi and Talmon [138]/American Chemical Society/CC BY 4.0.; (c) Jiang et al. [139]/with permission of Oxford University Press.

of the sample, and the sample becomes glassy without forming ice crystals by rapid freezing. In this state, the sample deteriorates much more slowly in order to facilitate the study of EV structure [136].

Some undeformed EVs were identified using Cryo-M by Bairamukov's team at the Petersburg Institute of Nuclear Physics (Figure 2.4a) [137]. Talmon et al. studied the appearance of EV through Cryo-electron microscopy. The darker area was irradiated longer than the lighter area, exposing the CNTs after etching away some of the CSA, showing clearly bundles of CNTs, forming a nematic liquid–crystalline phase. The bright spots are nanoparticles of the iron catalyst exposed by etching (Figure 2.4b) [138]. The morphological structure, size distribution, and membrane thickness of vesicles were distinctly determined via Cryo-M imaging by Jiang et al. (Figure 2.4c) [139].

2.2.2 Particle Size Detection

The morphology of EVs can be observed by electron microscopy, but the particle size distribution and overall concentration of all EVs in a sample cannot be represented. Particle size detection allows rapid and accurate analysis of the particle size distribution and concentration of EVs, providing strong evidence for the identification of EVs. Measuring the particle size distribution of EVs has long been an important part of EV characterization. However, since the size of EVs is only 30–200 nm, some special detection methods must be used to observe these invisible particles under the light microscope. The typical methods include nanoparticle tracking analysis (NTA) and dynamic light scattering (DLS).

2.2.2.1 Nanoparticle Tracking Analysis

NTA nanoparticle tracking analysis technology has been recognized as one of the means of exocrine characterization in the field of exocrine research. The principle is to track and analyze the Brownian motion of each particle and calculate the Fluid mechanics diameter and concentration of nanoparticles. Sample processing is simpler, more able to ensure the original state of exosomes, and detection speed is faster. NTA techniques can be used to determine the size and concentration of particles, and to measure the zeta potential; the higher the potential, the less likely it is that particles in solution will aggregate and precipitate [140]. Smaller and less-numerous particles are easily overlooked by traditional measurement methods. The NTA is capable of measuring particle sizes between 10 and 1000 nm in diameter.

The NTA tracks each particle in the image directly during the measurement process, so the NTA has a very high resolution for complex samples and can clearly distinguish between particles of different sizes, which is important for the detection of EVs. The NTA uses a measurement technique that eliminates the need for prior knowledge of the mass, refractive index, and hydrodynamic diameter of the particle material, allowing the instrument to skip the tedious preparatory work and move directly into the analysis, which not only saves a lot of time and speeds up the entire research process but also saves valuable manpower costs. Its unique concentration measurement technique provides reliable concentration data directly to researchers of EVs. Prior to NTA analysis of EVs in biological fluids, such as blood, it is essential to separate and purify the vesicles by centrifugation or other methods in order to remove lipoprotein particles, protein complexes, and other particles that may be similar in size to EVs and may exceed the number of EVs in the blood.

For example, Kenneth Gouin et al. used NTA to count and characterize the exocysts of different types of cardiomyocyte-derived cells in terms of particle size. The experimental results showed that the exocysts secreted by different cell donors were similar in size, and the particle size was mainly distributed between 70 and 90 nm (Figure 2.5a) [141]. Zhang Jinchao et al. developed a tumor antigen-carrying exosome (tDC-Exo) and obtained antibody-engineered exosomes (Exo-OVA-aCD3/aEGFR) by modifying anti-CD3 and anti-EFGR to improve tumor therapy. Particle size assay using NTA showed that the two types of exosomes—Exo-OVA with a diameter of about 93 nm and Exo-OVA-aCD3/aEGFR with a particle size of 102 nm (Figure 2.5b) [142]—can flexibly cross the tissue barrier and reach the target tissue more easily. Both animals and plants can secrete EVs; but plant-derived EVs are widely available compared with animal-derived EVs, and can be extracted and isolated in large quantities, which is more advantageous for tumor nanomedicine development and application. Liu et al. extracted EVs from plant leaves and evaluated the size distribution of leaf nano-EVs and small extracellular vesicles (sEVs) by NTA. The results showed that the size of most leaf nanovesicles was 218 nm, while the particle size of sEVs was 135 nm (Figure 2.5c) [143].

2.2.2.2 Dynamic Light Scattering

Like NTA, DLS is used to reflect the size and concentration of particles by using the scattered light produced by the Brownian motion of the particles. However, rather than using the scattered light to determine the diffusion coefficient of the particles, DLS uses fluctuations in the intensity of the scattered light to estimate the size of the particles. Unlike NTA, DLS requires a very small sample size, typically 70 μl. And it is easy to use because few parameters are required for optimization. DLS can be used to characterize the particle size of proteins, macromolecules, micelles, sugars, and nanoparticles. If the sample is a monodisperse system, the average effective diameter of the particles can be found. This measurement depends on the surface structure of the particles, the concentration of the particles, and the type of ions in the medium. DLS can also be used for stability studies, where the trend of particle

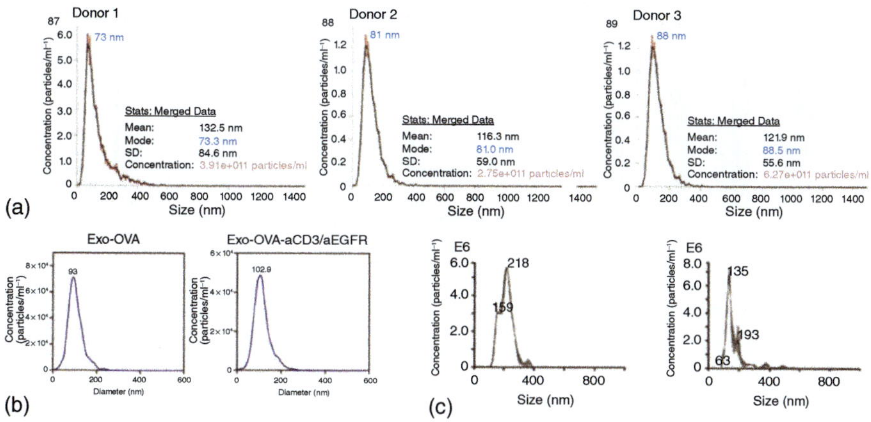

Figure 2.5 The size of EVs was identified by NTA. (a) Nanotracking analysis of EVs derived from three donor sources [141]; (b) NTA results of Exo-OVA and Exo-OVA-aCD3/aEGFR [142]; (c) Leaf nanovesicle and sEVs size distributions, as assessed by nanoparticle tracking analysis [143].

aggregation and sedimentation over time can be demonstrated by measuring the particle size distribution at different times. As particles agglomerate and sink, more particles with larger particle sizes become available. Similarly, DLS can be used to analyze the effect of temperature on stability.

The DLS technique has the advantages of short detection time and low cost; therefore, when the specific morphology of the nanoparticles is not required, the DLS method has advantages in terms of cost, time, ease of operation, and practical production. The DLS technique has the advantage of being unmatched by other methods in terms of cost, time, ease of operation, and specific practical production.

Although DLS has its advantages over NTA, the main disadvantages of DLS are its susceptibility to interference from the polydispersity of particles in the system and its low resolution when analyzing nonhomogeneous mixtures. For example, the intensity of the scattered light is positively correlated with the sixth power of the particle diameter, making it more difficult to detect scattered light from smaller particles. Therefore, when particles of different sizes are present in the mixture, the data generated by the scattered light tend to be biased toward the larger particle size, making the measured particle size somewhat larger than the true one. Furthermore, in order to obtain the characteristic particle peaks, it is necessary that the particle size between particles meets at least a threefold difference (e.g. 30 and 90 nm). At the same time, because the presence of particles is sensed optically, DLS can only confirm the number and diameter of the "vesicles" in the EV sample as a population but not whether the detected particles are EVs, which does not reflect the morphology and structure of the particles, and cannot distinguish EVs from other nanoparticles.

Lee et al. used DLS to detect the size of EVs secreted by MDA-MB-231 cells treated with Ac4ManNAz, Ac4GalNAz, and Ac4GlcNAz, and the results showed that there was no significant difference in the size of fluorescent-labeled cell-derived EVs treated with different azides compared with normal EVs (Figure 2.6a) [144].

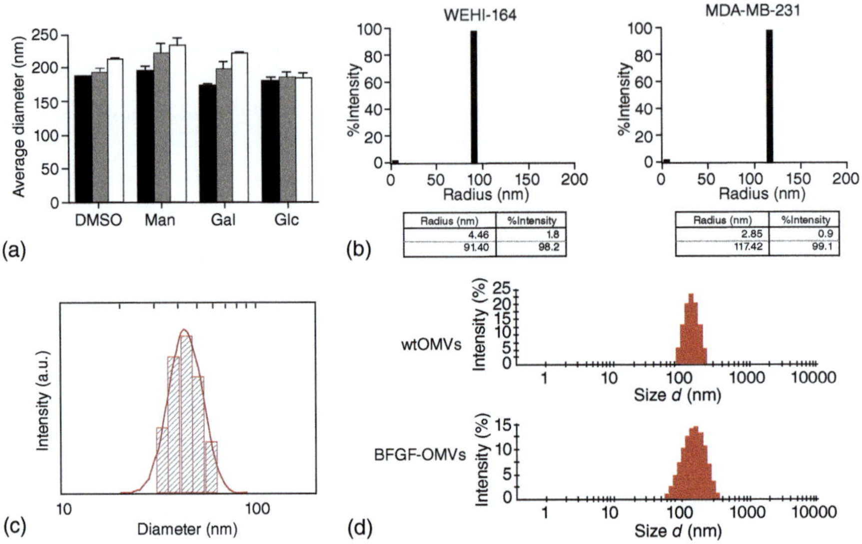

Figure 2.6 The size of EVs was identified by DLS. (a) the size of MDA-MB-231 cells derived exosomes was measured by Zetasizer Nano-ZS [144]; (b) A representative DLS result for the size distribution of exosomes [145]; (c, d) The particle size of vesicles of Gram negative bacteria at different temperatures [147].

Lyu et al. used DLS to analyze the radius of the main peak at which EVs from fibrosarcoma cell lines WEHI-164 and breast cancer cell lines MDA-MD-231 reached the maximum strength. The mean radius of EVs from WEHI-164 and MDA-MB-231 cells was 94.4 and 107.8 nm, respectively, indicating high homogeneity (Figure 2.6b) [145]. The outer membrane vesicles (OMVs) released by bacteria strongly activate the innate immune system and can be used as immune adjuvants. Therefore, OMVs are ideal vaccine nanocarriers. Sarra et al. used DLS method to study and analyze vesicles from Gram-negative bacteria at different temperatures [146]. The weighted distribution of dynamic light scattering intensity of *Escherichia coli* OMV at 37 °C showed that the average hydrodynamic diameter of vesicles was 48 ± 3 nm (FWHM = 24 ± 2 nm). The mean diameters of OMV grown at 27 and 20 °C were 37 ± 4 nm (FWHM = 32 ± 2 nm) and 24 ± 2 nm (FWHM = 20 ± 3 nm), respectively. Notably, DLS analysis showed that vesicle size decreased with decreasing growth temperature. Huang and his colleagues extracted OMV (wtOMVs) naturally secreted by *E. coli* DH5α and OMVs (BfGF-OMVs) modified with basic fibroblast growth factor (BFGF) molecules for size comparison. DLS results showed that the average particle diameters of wtOMVs and BFGF-OMVs were 177.5 and 166.9 nm, respectively, and the PdI values were 0.245 and 0.358, respectively (Figure 2.6c,d). These results indicate that the homogeneity of OMVs is affected, and the particle size range is enlarged after BFGF modification [147].

2.2.3 Surface Protein Assay

The Tetraspanin family (CD9, CD63, and CD81), the heat shock protein family (HSP60, HSP70, HSPA5, CCT2, and HSP90), and some cell-specific proteins, including A33 (from colon epithelial cells), MHC-II (from antigen presenting cells), CD86 (from antigen presenting cells), and milk Lectin (from immature dendritic cells), are protein substances rich in the membrane of exosomes and participate in exosomes. CD63, CD9, and CD81, as well as TSG101, ESCRT, and ALIX, are the most commonly used markers to detect EVs [148]. These marker proteins are mainly involved in the process of vesicle formation and secretion. They are produced in cells located in the vicinity of intracellular membrane structures, such as Tetraspanin (CD9, CD63, and CD81), which are directly involved in the sorting of EV contents. TSG101 is a protein associated with the ESCRT complex, which, in turn, is a key driver of membrane formation and rupture. ALIX is directly involved in the process of membrane vesicle formation by cutting off from the plasma membrane during the formation of separate membrane structures.

There are three main methods to detect surface marker proteins: protein immunoblotting, nanofluidics and enzyme-linked immunosorbent assay.

2.2.3.1 Protein Immunoblotting Method

The basic principle of western blot (WB) is to obtain information about the expression of a particular protein in the cells or tissues being analyzed by analyzing the position and depth of the coloring by using antibodies specific to the cells or biological tissues treated by gel electrophoresis. The WB method is similar to Southern Blot or Northern Blot hybridization, but uses polyacrylamide gel electrophoresis, where the protein is detected, the "probe" is an antibody, and the "color development" is done with a labelled secondary antibody. The polyacrylamide gel electrophoresis (PAGE)-separated protein sample is transferred to a solid-phase carrier (e.g. nitrocellulose film), and adsorbs the protein in a noncovalent bond, which can keep the type of peptide. The protein or peptide on the solid-phase carrier is used as an antigen, which reacts with a corresponding antibody, and then with an enzyme or isotopically labelled secondary antibody, which is used to detect the protein component expressed in the specific target gene separated by electrophoresis, after substrate development or radioautography. The results of EVs extracted by different methods vary slightly with this method.

As an important component in profiling the function of EVs, protein markers are increasingly required for their sensitivity, specificity, and clinical validity. In the recent years, new technological platforms with different principles have been gradually applied to protein detection in EV and are gradually being used for clinical disease diagnosis, condition monitoring, and prognostic assessment. The identification of specific proteins in EVs by WB can indicate the presence of EV components. However, these marker proteins have been found to be enriched in EV so far; but they are not unique to EV. For example, CD63 and ALIX are also present in a large number cells, cell debris, lysosomes, and intracellular bodies. Therefore, protein immunoblotting can only confirm the presence of EV components, but not the quantity and integrity of EVs.

Que et al. characterized tumor exosomes using a WB method to validate key differentially expressed proteins. The levels of EGF, p-mTOR, p-STAT1, STAT3, p-STAT3, PRKCA, p-PRKCA, Akt1/2/3, and p-Akt were detected by WB assay 24 hours after drug action on CTC-TJH-01 cells, as shown in Figure 2.7a. Wang et al. from

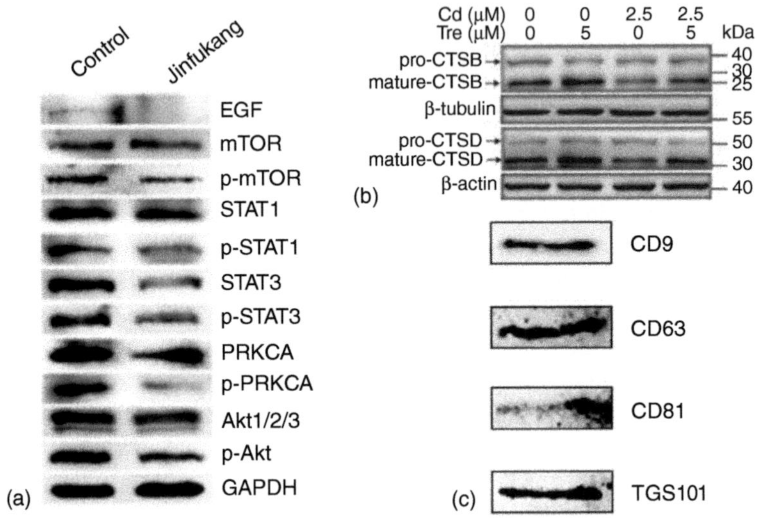

Figure 2.7 WB result of representative exosomal proteins. Source: (a) Que et al. [149]/Elsevier/CC BY 4.0.; (b) Wang et al. [150]/Springer Nature/CC BY 4.0.; (c) Domenis et al. [151]/Mediators of Inflammation/CC BY 4.0.

Shandong Agricultural University used flow cytometry and immunoblot analysis to identify the apoptotic vesicles produced by mouse kidney cells from a morphological perspective, while the process of apoptosis was documented by observing the immunoblot analysis of autophagy marker proteins [149]. The final results showed that alginate (Tre) significantly restored the cadmium (Cd)-promoting effect on autophagic vesicle-lysosome fusion, thereby enhancing autophagic degradation in cadmium (Cd)-exposed primary rat proximal renal tubule (rPT) cells. Histone proteases are the main lysosomal proteases involved in autophagic degradation, of which CTSB and CTSD are two abundant lysosomal proteases. As acidification is required for the maturation (activation) of histone proteases, the change in pH ultimately leads to much reduced protein degradation. As shown in Figure 2.7b, Cd impaired the maturation of CTSB and CTSD in rPT cells, and co-treatment with Tre resulted in a significant restoration of CTSB and CTSD maturation. Based on these results, it can be concluded that Tre restored the impaired autophagic flux in Cd-treated rPT cells, in part, by enhancing lysosomal function. Wang et al. isolated EVs from inflamed SF (patient synovial fluid) by polymer precipitation and quantified them by Exocet kit and nanoparticle tracking analysis [150]; it was followed by the authors' use of protein blotting used to identify EVs isolated from inflamed patient synovial fluid, demonstrating that SF-derived exosomes stimulated the release of several inflammatory cytokines from M1 macrophages. The authors lysed SF-derived exosomes and separated 2 μg of protein on 15% SDS-PAGE gels under reducing conditions. The gels were coated on nitrocellulose membranes and stained with antibodies against the exosomal markers CD9, CD63, CD81, and TSG101, and the results are shown in Figure 2.7c [151].

2.2.3.2 Nano-Flow Cytometry

Nano-flow cytometry (nFCM) can detect the scattered light signal of EVs down to 30 nm, thus enabling the detection of individual phaeohemoglobin molecules and completely distinguishing them from the background. The nFCM can establish a standard method for multiparameter quantitative analysis of EVs at the single-particle level, which provides a solid technical basis for promoting the wide application of EVs in the diagnosis and treatment of clinical diseases [152].

One of the features of the nFCM assay is its high sensitivity, allowing the detection of scattered light from low-refractive-index nanoparticles at 24 nm and the fluorescence of individual phycoerythrins. nFCM is suitable for a wide range of applications, both for unlabelled and labelled samples. nFCM can simultaneously detect multiple parameters of individual nanoparticles, measuring particle size by scattered light and revealing multiple biochemical properties of nanoparticles by multicolor fluorescence. nFCM can be used for quantitative analysis, and the resolution of nanoparticle particle size characterization is comparable to that of cryo-transmission electron microscopy. nFCM sample volumes as low as 10 μl and actual consumption as low as 1 μl are particularly important for the determination of valuable biological samples; and nFCM acquisition is fast, with highly representative statistical distribution characteristics of particle traits obtained in less than one minute. Compared to protein immunoblotting, nFCM has the advantage of requiring a smaller sample size for the assay, making it easier to quantify the proportion of different subpopulations. At the same time, nFCM offers high throughput, and the ability to accurately measure diameter and concentration.

Busatto et al. investigated the dose-dependent uptake of nanoscale human serum exosomes by tumor-bearing mice and human cells by increasing concentrations of exosomes, and grading exosomes by purity, titrated by colorimetric nano-plasma (CONAN) assay followed by cell flow cytofluorimetric analysis. The results showed that serum exosomes from healthy humans were internalized by mouse and human cells in a specific dose-dependent manner (Figure 2.8a) [153]. Wang et al. from Liaoning University analyzed apoptotic vesicles in cancer cells using flow cytometry and found that bexarotene nanocrystals significantly enhanced in-vitro cytotoxicity, and induced G1 cycle arrest and apoptosis in A549 cells (Figure 2.8b) [154]. Morales-Kastresana et al. used nanoFACS—a high-resolution flow cytometer to better discriminate EVs using light scattering and fluorescence parameters and sample enumeration to evaluate various markers. Efficient characterization of EVs by nanoFACS has paved the way for further studies on the role of EVs in health and disease, paving the way for further studies on the role of EVs in health and disease (Figure 2.8c) [155].

2.2.3.3 Enzyme-Linked Immunosorbent Assay

Enzyme-linked immunosorbent assay (ELISA) refers to a qualitative and quantitative assay that binds a soluble antigen or antibody to a solid-phase carrier, such as polystyrene, and uses the specificity of antigen–antibody binding for the immune reaction. It is one of the most applied immunoenzymatic techniques. ELISA kits

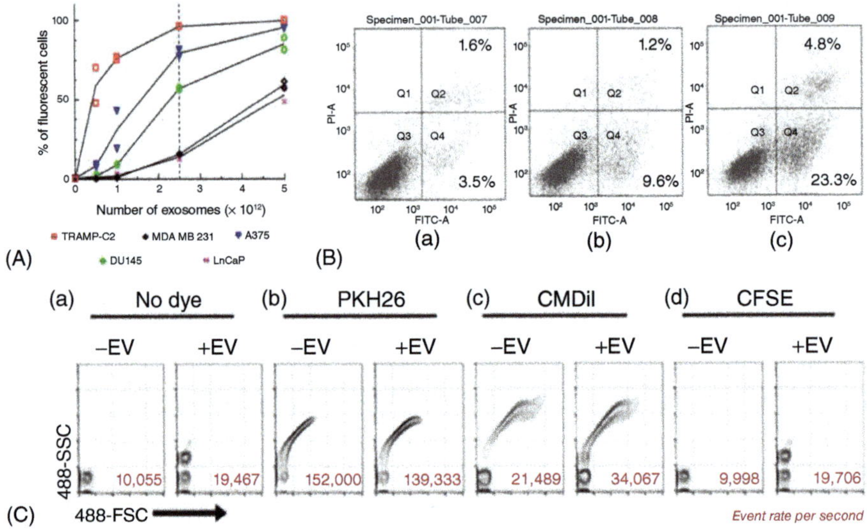

Figure 2.8 nFCM result of EVs. (a) Dose-dependent uptake profiles of human serum exosomes by tumor derived murine and human cells [153]; (b) The Cell apoptosis, morphological changes and early apoptosis rate of human lung adenocarcinoma epithelial cell line A549 treated with Bex-Sol and NC-Bex for 24 h [154]; (c) EV label suitability analysis by nanoFACS HR-FCM and NTA. nanoFACS analysis of light scattering pattern and event rate [155].

have become a common diagnostic tool in drug and phytopathology studies, ideal for detecting target antibodies/antigens in tissue extracts, serum and cell cultures, and an important quality control tool.

The ExoELISA kit is a quantitative assay for human EVs based on ELISA technology, which obtains EV concentration results based on a standard curve of known EV numbers, specifically binding proteins including CD63, CD9, and CD81. As protein expression varies between different types of body fluids and cell lines, samples often need to be tested multiple times, resulting in the need for larger sample volumes and variation between test batches.

ELISA is used for the digital identification of target exosomes using droplet microfluidics. The exosomes were immobilized on magnetic microbeads through sandwich ELISA complexes tagged with an enzymatic reporter that produces a fluorescent signal. The constructed beads were further isolated and encapsulated into a sufficient number of droplets to ensure that only a single bead was encapsulated in a droplet. Droplet-based single-exosome-counting enzyme-linked immunoassay (droplet digital ExoELISA) approaches enable absolute counting of cancer-specific exosomes to achieve unprecedented accuracy. It was able to achieve a limit of detection (LOD) down to 10 enzyme-labeled exosome complexes per microliter (~10^{-17} M). We demonstrated the application of the droplet digital ExoELISA platform in quantitative detection of exosomes in plasma samples directly from breast cancer patients (Figure 2.9) [156].

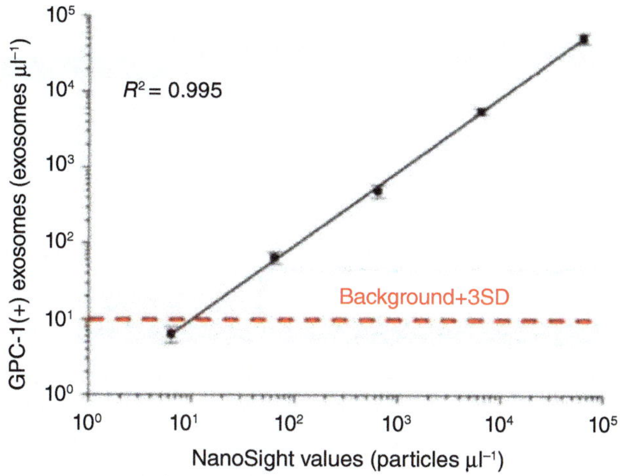

Figure 2.9 Droplet digital ExoELISA 3 calibration results showing that the dynamic range of the captured exosomes spans 5 orders of 4 magnitude [156].

2.2.4 Other Methods

Many new and emerging techniques for exosome isolation and identification (such as single EV analysis [SEA], tunable resistive pulse sensing, micronuclear magnetic resonance, electrodialysis, immunoaffinity capture followed by elution) have also been put into practice, taking into account the heterogeneity of EVs and the nuances of their properties. Some of these methods are described in detail below.

2.2.4.1 Tunable Resistive Pulse Sensing

Tunable resistive pulse sensing (TRPS) is a real-time nanoparticle analysis technique capable of measuring nanoparticles in the size range from 60 nm to 2 μm, which can be used instead of nFCM to measure the concentration and size distribution of EVs. This technology is based on the nanoscale Coulter principle. It detects transient changes in ion flow caused by the transport of vesicles through tunable nanopores in polyurethane membranes. When a certain voltage is applied to the top and bottom of the nanopore filled with electrolyte, an ionic current is generated inside the nanopore. As the particles pass through the nanopores, they displace a certain amount of electrolyte to increase the resistance and then generate a resistance pulse signal. The magnitude of this pulse signal is proportional to the volume of the particles, and detection of particle size and concentration is achieved by using calibrated particles of known size and concentration.

TRPs are currently the most powerful particle analysis system in the determination and analysis of nano- and submicron particles, and can be used to measure the true size distribution (the concentration of particles in a specific size range) based on particle-by-particle surface charge determination. TRPS is mainly used in EVs, nanomedicine, virus, and vaccine characterization, etc.

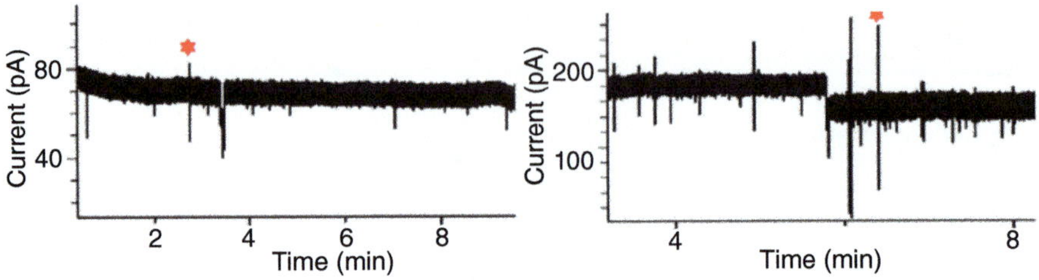

Figure 2.10 ERP current–time recordings obtained with a CNP positioned near MCF-10A and MDA-MB-231 cell surfaces [157].

Electrochemical resistance pulse (ERP) sensing enables the direct detection of individual EVs released from specific cells, and the analysis of reactive oxygen and nitrogen species in such vesicles. Here, we demonstrate the applicability of ERP sensing in distinguishing between nontransformed and cancerous breast cell lines and breast cancer cell lines with different metastatic potential. Another application of ERP sensing is the real-time monitoring of changes in individual cells induced by chemical agents. As shown below, by studying nontransformed human breast cells (MCF-10A) and metastatic cancer cells (MDA-MB-231)s' EVs, Jia et al. found that EVs released from MDA-MB-231 cells produced Faraday current peaks, whereas the ERP recordings obtained from MCF-10A cells contained few anodic peaks and were relatively small in magnitude (Figure 2.10) [157].

2.2.4.2 Single EV Analysis Technique

The SEA technique allows for stable, multicomponent protein biomarker measurements in a single vesicle. In this method, EVs are immobilized in a microfluidic cavity by immunostaining and imaging. While the vesicles are immobilized on the wafer surface, the signal-to-noise ratio of each vesicle is typically much higher than that of vesicles in the solution or flow state. The SEA technique may be a powerful tool for studying a variety of EV types by providing a rich dataset of biomarker expression heterogeneity, marker composition, and EV subpopulation identification.

The SEA technique enables stable, multicomponent protein biomarker measurements in a single vesicle. As shown in the figure below, EVs are immobilized in a microfluidic chamber, immunostained, and imaged. When the vesicles are immobilized on the wafer surface, the signal-to-noise ratio of each vesicle is typically much higher than that of vesicles in solution or in the flow state. The authors further adapted the image cycling process previously used for multiplexed cell and tissue analysis to complement the nanoscale size of EVs. Specifically, the authors used EVs derived from three homozygous glioblastoma multiforme (GBM) cell lines. EVs were stained for three protein markers at a time and then subjected to four rounds of cycling for imaging. The results showed high heterogeneity in the biomarkers of EVs. The data were then visualized using t-distributed random neighbor embedding

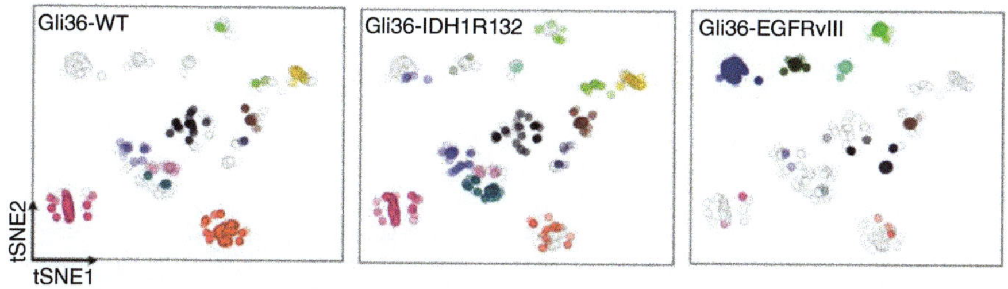

Figure 2.11 2-dimensional tSNE mapping of individual EVs analyzed for proteinmarkers [158].

(tSNE); unsupervised clustering revealed the presence of potential EV subgroups. sEA technology may be a powerful tool for studying various EV types, providing a rich dataset on biomarker expression heterogeneity, marker composition, and identification of EV subgroups (Figure 2.11) [158].

2.2.4.3 Micronuclear Magnetic Resonance

Micronuclear magnetic resonance (μNMR) is mainly based on magnetic nanoparticle properties. Due to the natural lack of ferromagnetic background in most biological materials, this sensing is almost immune to interference from other biological samples in the same system. Therefore, even optically turbid samples are transparent to magnetic fields. When target molecules are targeted by specific magnetite nanoparticles (MNPs), they form a strong contrast with the natural biological background. Direct measurement of EV using micro-NMR on a microfluidic chip allows determination of the abundance of EV biomarkers. In nuclear magnetic resonance (NMR)-based magnetic detection, MNPs are placed in an NMR magnetic field, which generates a local magnetic field that alters the lateral relaxation rate of surrounding water molecules and amplifies the analyzed signal. As a result, NMR reduces the sample handling process, improves the detection sensitivity, and has been developed for several point-of-care applications. The technique is more sensitive than conventional protein techniques, and the μNMR system is 103 times more sensitive than WB and Elisa.

There are difficulties in making NMR applicable to EV detection, because EVs tumor cells are one to two orders of magnitude smaller. To bridge this size gap, Shao et al. developed a new analytical technique specifically for EV detection and protein analysis using two-step bioorthogonal click chemistry to label EV with MNP. As shown below, the use of a small molecule labeling strategy did not significantly increase the size of the antibody or MNP, thereby increasing the efficiency of retaining unbound antibodies and MNP for targeting vesicles. Targeted EVs were then measured directly on the microchip using microfluidic micronuclear magnetic resonance (μNMR) to determine the abundance of EV biomarkers (Figure 2.12) [159].

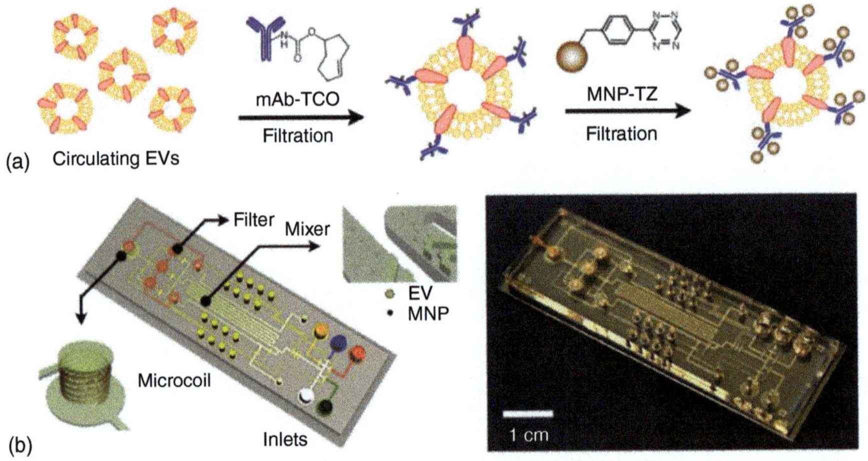

Figure 2.12 The microfluidic system for on-chip detection of circulating EVs is designed to detect MNP-targeted vesicles, concentrate MNP-tagged vesicles (while removing unbound MNPs), and provide in-line NMR detection [159]. Source: (b) Salmon et al. 2022/John Wiley & Sons/CC BY

References

1 van Niel, G., D'Angelo, G., and Raposo, G. (2018). Shedding light on the cell biology of extracellular vesicles. *Nat. Rev. Mol. Cell Biol.* 19 (4): 213–228.

2 Zhang, R., Wang, L., Zhao, J., and Guo, S. (2018). Effects of sodium alginate on the composition, morphology, and electrochemical properties of electrospun carbon nanofibers as electrodes for supercapacitors. *ACS Sustainable Chem. Eng.* 7 (1): 632–640.

3 Broad, K., Walker, S.A., Davidovich, I. et al. (2023). Unraveling multilayered extracellular vesicles: Speculation on cause. *J. Extracell. Vesicles* 12 (2): e12309.

4 Herrmann, I.K., Wood, M.J.A., and Fuhrmann, G. (2021). Extracellular vesicles as a next-generation drug delivery platform. *Nat. Nanotechnol.* 16 (7): 748–759.

5 Shah, R., Patel, T., and Freedman, J.E. (2018). Circulating extracellular vesicles in human disease. *N. Engl. J. Med.* 379 (10): 958–966.

6 Akers, J.C., Gonda, D., Kim, R. et al. (2013). Biogenesis of extracellular vesicles (EV): exosomes, microvesicles, retrovirus-like vesicles, and apoptotic bodies. *J. Neurooncol.* 113 (1): 1–11.

7 Shao, H., Im, H., Castro, C.M. et al. (2018). New technologies for analysis of extracellular vesicles. *Chem. Rev.* 118 (4): 1917–1950.

8 Morello, M., Minciacchi, V.R., de Candia, P. et al. (2013). Large oncosomes mediate intercellular transfer of functional microRNA. *Cell Cycle* 12 (22): 3526–3536.

9 Hessvik, N.P. and Llorente, A. (2018). Current knowledge on exosome biogenesis and release. *Cell. Mol. Life Sci.* 75 (2): 193–208.

10 Kalluri, R. and LeBleu, V.S. (2020). The biology, function, and biomedical applications of exosomes. *Science* 367 (6478).

11 Xu, Y., Feng, K., Zhao, H. et al. (2022). Tumor-derived extracellular vesicles as messengers of natural products in cancer treatment. *Theranostics* 12 (4): 1683–1714.

12 Teng, F. and Fussenegger, M. (2020). Shedding light on extracellular vesicle biogenesis and bioengineering. *Adv. Sci.* 8 (1): 2003505.

13 Henne, W.M., Buchkovich, N.J., and Emr, S.D. (2011). The ESCRT pathway. *Dev. Cell* 21 (1): 77–91.

14 Juan, T. and Furthauer, M. (2018). Biogenesis and function of ESCRT-dependent extracellular vesicles. *Semin. Cell Dev. Biol.* 74: 66–77.

15 Schoneberg, J., Lee, I.H., Iwasa, J.H., and Hurley, J.H. (2017). Reverse-topology membrane scission by the ESCRT proteins. *Nat. Rev. Mol. Cell Biol.* 18 (1): 5–17.

16 Williams, R.L. and Urbe, S. (2007). The emerging shape of the ESCRT machinery. *Nat. Rev. Mol. Cell Biol.* 8 (5): 355–368.

17 Raiborg, C. and Stenmark, H. (2009). The ESCRT machinery in endosomal sorting of ubiquitylated membrane proteins. *Nature* 458 (7237): 445–452.

18 Yin, M., Li, Z., Liu, Z. et al. (2012). Magnetic self-assembled zeolite clusters for sensitive detection and rapid removal of mercury (II). *ACS Appl. Mater. Interfaces* 4 (1): 431–437.

19 Wurmser, A.E. and Emr, S.D. (1998). Phosphoinositide signaling and turnover: PtdIns(3)P, a regulator of membrane traffic, is transported to the vacuole and degraded by a process that requires lumenal vacuolar hydrolase activities. *EMBO J.* 17 (17): 4930–4942.

20 Raiborg, C. (2001). Hrs recruits clathrin to early endosomes. *EMBO J.* 20 (17): 5008–5021.

21 Tamai, K., Tanaka, N., Nakano, T. et al. (2010). Exosome secretion of dendritic cells is regulated by Hrs, an ESCRT-0 protein. *Biochem. Biophys. Res. Commun.* 399 (3): 384–390.

22 Raiborg, C., Bache, K.G., Gillooly, D.J. et al. (2002). Hrs sorts ubiquitinated proteins into clathrin-coated microdomains of early endosomes. *Nat. Cell Biol.* 4 (5): 394–398.

23 Bilodeau, P.S., Urbanowski, J.L., Winistorfer, S.C., and Piper, R.C. (2002). The Vps27p Hse1p complex binds ubiquitin and mediates endosomal protein sorting. *Nat. Cell Biol.* 4 (7): 534–539.

24 Ju, E., Li, Z., Li, M. et al. (2013). Functional polypyrrole–silica composites as photothermal agents for targeted killing of bacteria. *Chem. Commun.* 49 (79): 9048–9050.

25 Bache, K.G., Brech, A., Mehlum, A., and Stenmark, H. (2003). Hrs regulates multivesicular body formation via ESCRT recruitment to endosomes. *J. Cell Biol.* 162 (3): 435–442.

26 Bilodeau, P.S., Winistorfer, S.C., Kearney, W.R. et al. (2003). Vps27-Hse1 and ESCRT-I complexes cooperate to increase efficiency of sorting ubiquitinated proteins at the endosome. *J. Cell Biol.* 163 (2): 237–243.

27 Teo, H., Gill, D.J., Sun, J. et al. (2006). ESCRT-I Core and ESCRT-II GLUE Domain Structures Reveal Role for GLUE in Linking to ESCRT-I and Membranes. *Cell* 125 (1): 99–111.

28 Alam, S.L., Langelier, C., Whitby, F.G. et al. (2006). Structural basis for ubiquitin recognition by the human ESCRT-II EAP45 GLUE domain. *Nat. Struct. Mol. Biol.* 13 (11): 1029–1030.

29 Slagsvold, T., Aasland, R., Hirano, S. et al. (2005). Eap45 in mammalian ESCRT-II binds ubiquitin via a phosphoinositide-interacting GLUE domain. *J. Biol. Chem.* 280 (20): 19600–19606.

30 Gill, D.J., Teo, H., Sun, J. et al. (2007). Structural insight into the ESCRT-I/-II link and its role in MVB trafficking. *EMBO J.* 26 (2): 600–612.

31 Wollert, T. and Hurley, J.H. (2010). Molecular mechanism of multivesicular body biogenesis by ESCRT complexes. *Nature* 464 (7290): 864–869.

32 Teo, H., Perisic, O., González, B., and Williams, R.L. (2004). ESCRT-II, an endosome-associated complex required for protein sorting. *Dev. Cell* 7 (4): 559–569.

33 Saksena, S., Wahlman, J., Teis, D. et al. (2009). Functional reconstitution of ESCRT-III assembly and disassembly. *Cell* 136 (1): 97–109.

34 Yang, X., Li, Z., Ju, E. et al. (2014). Reduced graphene oxide functionalized with a luminescent rare-earth complex for the tracking and photothermal killing of drug-resistant bacteria. *Chem. Eur. J.* 20 (2): 394–398.

35 Babst, M. (1997). Endosomal transport function in yeast requires a novel AAA-type ATPase, Vps4p. *EMBO J.* 16 (8): 1820–1831.

36 Wollert, T., Wunder, C., Lippincott-Schwartz, J., and Hurley, J.H. (2009). Membrane scission by the ESCRT-III complex. *Nature* 458 (7235): 172–177.

37 Baietti, M.F., Zhang, Z., Mortier, E. et al. (2012). Syndecan–syntenin–ALIX regulates the biogenesis of exosomes. *Nat. Cell Biol.* 14 (7): 677–685.

38 Roucourt, B., Meeussen, S., Bao, J. et al. (2015). Heparanase activates the syndecan-syntenin-ALIX exosome pathway. *Cell Res.* 25 (4): 412–428.

39 Ghossoub, R., Lembo, F., Rubio, A. et al. (2014). Syntenin-ALIX exosome biogenesis and budding into multivesicular bodies are controlled by ARF6 and PLD2. *Nat. Commun.* 5 (1).

40 Imjeti, N.S., Menck, K., Egea-Jimenez, A.L. et al. (2017). Syntenin mediates SRC function in exosomal cell-to-cell communication. *Proc. Natl. Acad. Sci. U.S.A.* 114 (47): 12495–12500.

41 Stefani, F., Zhang, L., Taylor, S. et al. (2011). UBAP1 is a component of an endosome-specific ESCRT-I complex that is essential for MVB sorting. *Curr. Biol.* 21 (14): 1245–1250.

42 Ali, N., Zhang, L., Taylor, S. et al. (2013). Recruitment of UBPY and ESCRT exchange drive HD-PTP-dependent sorting of EGFR to the MVB. *Curr. Biol.* 23 (6): 453–461.

43 Ge, K., Zhang, C., Sun, W. et al. (2016). Up-conversion Y_2O_3: Yb^{3+}, Er^{3+} hollow spherical drug carrier with improved degradability for cancer treatment. *ACS Appl. Mater. Interfaces* 8 (38): 25078–25086.

44 Dores, M.R., Paing, M.M., Lin, H. et al. (2012). AP-3 regulates PAR1 ubiquitin-independent MVB/lysosomal sorting via an ALIX-mediated pathway. *Mol. Biol. Cell* 23 (18): 3612–3623.

45 Dores, M.R., Chen, B., Lin, H. et al. (2012). ALIX binds a YPX3L motif of the GPCR PAR1 and mediates ubiquitin-independent ESCRT-III/MVB sorting. *J. Cell Biol.* 197 (3): 407–419.

46 Dong, K., Li, Z., Sun, H. et al. (2017). Pathogen-mimicking nanocomplexes: self-stimulating oxidative stress in tumor microenvironment for chemo-immunotherapy. *Mater. Today* 20 (7): 346–353.

47 Duan, F., Feng, X., Jin, Y. et al. (2017). Metal–carbenicillin framework-based nanoantibiotics with enhanced penetration and highly efficient inhibition of MRSA. *Biomaterials* 144: 155–165.

48 Stuffers, S., Sem Wegner, C., Stenmark, H., and Brech, A. (2009). Multivesicular endosome biogenesis in the absence of ESCRTs. *Traffic* 10 (7): 925–937.

49 Ren, H., Chen, S., Jin, Y. et al. (2017). A traceable and bone-targeted nanoassembly based on defect-related luminescent mesoporous silica for enhanced osteogenic differentiation. *J. Mater. Chem. B* 5 (8): 1585–1593.

50 Dikic, I. (2004). ALIX-ing phospholipids with endosome biogenesis. *Bioessays* 26 (6): 604–607.

51 Chairoungdua, A., Smith, D.L., Pochard, P. et al. (2010). Exosome release of beta-catenin: a novel mechanism that antagonizes Wnt signaling. *J. Cell Biol.* 190 (6): 1079–1091.

52 Gulbins, E. and Kolesnick, R. (2003). Raft ceramide in molecular medicine. *Oncogene* 22 (45): 7070–7077.

53 Cheng, Q., Li, X., Wang, Y. et al. (2018). The ceramide pathway is involved in the survival, apoptosis and exosome functions of human multiple myeloma cells in vitro. *Acta Pharmacol. Sin.* 39 (4): 561–568.

54 van Niel, G., Charrin, S., Simoes, S. et al. (2011). The tetraspanin CD63 regulates ESCRT-independent and -dependent endosomal sorting during melanogenesis. *Dev. Cell* 21 (4): 708–721.

55 Hemler, M.E. (2003). Tetraspanin proteins mediate cellular penetration, invasion, and fusion events and define a novel type of membrane microdomain. *Annu. Rev. Cell Dev. Biol.* 19: 397–422.

56 Hemler, M.E. (2014). Tetraspanin proteins promote multiple cancer stages. *Nat. Rev. Cancer* 14 (1): 49–60.

57 Hurwitz, S.N., Conlon, M.M., Rider, M.A. et al. (2016). Nanoparticle analysis sheds budding insights into genetic drivers of extracellular vesicle biogenesis. *J. Extracell. Vesicles* 5: 31295.

58 van Niel, G., Bergam, P., Di Cicco, A. et al. (2015). Apolipoprotein E regulates amyloid formation within endosomes of pigment cells. *Cell Rep.* 13 (1): 43–51.

59 Gauthier, S.A., Perez-Gonzalez, R., Sharma, A. et al. (2017). Enhanced exosome secretion in down syndrome brain - a protective mechanism to alleviate neuronal endosomal abnormalities. *Acta Neuropathol. Commun.* 5 (1): 65.

60 Böker, K.O., Lemus-Diaz, N., Rinaldi Ferreira, R. et al. (2018). The impact of the CD9 tetraspanin on lentivirus infectivity and exosome secretion. *Mol. Ther.* 26 (2): 634–647.

61 Perez-Hernandez, D., Gutierrez-Vazquez, C., Jorge, I. et al. (2013). The intracellular interactome of tetraspanin-enriched microdomains reveals their function

as sorting machineries toward exosomes. *J. Biolumin. Chemilumin.* 288 (17): 11649–11661.
62 Villarroya-Beltri, C., Baixauli, F., Mittelbrunn, M. et al. (2016). ISGylation controls exosome secretion by promoting lysosomal degradation of MVB proteins. *Nat. Commun.* 7: 13588.
63 Wei, H., Chen, Q., Lin, L. et al. (2021). Regulation of exosome production and cargo sorting. *Int. J. Biol. Sci.* 17 (1): 163–177.
64 Jin, Y., Ma, L., Zhang, W. et al. (2022). Extracellular signals regulate the biogenesis of extracellular vesicles. *Biol. Res.* 55 (1): 35.
65 Colombo, M., Raposo, G., and Thery, C. (2014). Biogenesis, secretion, and intercellular interactions of exosomes and other extracellular vesicles. *Annu. Rev. Cell Dev. Biol.* 30: 255–289.
66 Ostrowski, M., Carmo, N.B., Krumeich, S. et al. (2010). Rab27a and Rab27b control different steps of the exosome secretion pathway. *Nat. Cell Biol.* 12 (1): 19–30; pp. 1–13.
67 Fei, X., Li, Z., Yang, D. et al. (2021). Neddylation of Coro1a determines the fate of multivesicular bodies and biogenesis of extracellular vesicles. *J. Extracell. Vesicles* 10 (12): e12153.
68 Savina, A., Fader, C.M., Damiani, M.T., and Colombo, M.I. (2005). Rab11 promotes docking and fusion of multivesicular bodies in a calcium-dependent manner. *Traffic* 6 (2): 131–143.
69 Bobrie, A., Krumeich, S., Reyal, F. et al. (2012). Rab27a supports exosome-dependent and -independent mechanisms that modify the tumor microenvironment and can promote tumor progression. *Cancer Res.* 72 (19): 4920–4930.
70 Peinado, H., Aleckovic, M., Lavotshkin, S. et al. (2012). Melanoma exosomes educate bone marrow progenitor cells toward a pro-metastatic phenotype through MET. *Nat. Med.* 18 (6): 883–891.
71 Hoshino, D., Kirkbride, K.C., Costello, K. et al. (2013). Exosome secretion is enhanced by invadopodia and drives invasive behavior. *Cell Rep.* 5 (5): 1159–1168.
72 Bobrie, A., Colombo, M., Krumeich, S. et al. (2012). Diverse subpopulations of vesicles secreted by different intracellular mechanisms are present in exosome preparations obtained by differential ultracentrifugation. *J. Extracell. Vesicles* 1.
73 Wei, D., Zhan, W., Gao, Y. et al. (2021). RAB31 marks and controls an ESCRT-independent exosome pathway. *Cell Res.* 31 (2): 157–177.
74 Hsu, C., Morohashi, Y., Yoshimura, S. et al. (2010). Regulation of exosome secretion by Rab35 and its GTPase-activating proteins TBC1D10A-C. *J. Cell Biol.* 189 (2): 223–232.
75 Jahn, R. and Scheller, R.H. (2006). SNAREs – engines for membrane fusion. *Nat. Rev. Mol. Cell Biol.* 7 (9): 631–643.
76 Liang, H., Li, Z., Ren, Z. et al. (2020). Light-triggered NO-releasing nanoparticles for treating mice with liver fibrosis. *Nano Res.* 13 (8): 2197–2202.
77 Li, H., Lei, W., Wu, J. et al. (2018). An upconverting nanotheranostic agent activated by hypoxia combined with NIR irradiation for selective hypoxia imaging and tumour therapy. *J. Mater. Chem. B* 6 (18): 2747–2757.

78 Fan, M., Han, Y., Gao, S. et al. (2020). Ultrasmall gold nanoparticles in cancer diagnosis and therapy. *Theranostics* 10 (11): 4944–4957.

79 Martens, S. and McMahon, H.T. (2008). Mechanisms of membrane fusion: disparate players and common principles. *Nat. Rev. Mol. Cell Biol.* 9 (7): 543–556.

80 Fader, C.M., Sanchez, D.G., Mestre, M.B., and Colombo, M.I. (2009). TI-VAMP/VAMP7 and VAMP3/cellubrevin: two v-SNARE proteins involved in specific steps of the autophagy/multivesicular body pathways. *Biochim. Biophys. Acta* 1793 (12): 1901–1916.

81 Rao, S.K., Huynh, C., Proux-Gillardeaux, V. et al. (2004). Identification of SNAREs involved in synaptotagmin VII-regulated lysosomal exocytosis. *J. Biolumin. Chemilumin.* 279 (19): 20471–20479.

82 Proux-Gillardeaux, V., Raposo, G., Irinopoulou, T., and Galli, T. (2007). Expression of the Longin domain of TI-VAMP impairs lysosomal secretion and epithelial cell migration. *Biol. Cell* 99 (5): 261–271.

83 Koles, K., Nunnari, J., Korkut, C. et al. (2012). Mechanism of evenness interrupted (Evi)-exosome release at synaptic boutons. *J. Biolumin. Chemilumin.* 287 (20): 16820–16834.

84 Gross, J.C., Chaudhary, V., Bartscherer, K., and Boutros, M. (2012). Active Wnt proteins are secreted on exosomes. *Nat. Cell Biol.* 14 (10): 1036–1045.

85 Zhu, D., Li, Z., Huang, K. et al. (2021). Minimally invasive delivery of therapeutic agents by hydrogel injection into the pericardial cavity for cardiac repair. *Nat. Commun.* 12 (1): 1–10.

86 Hyenne, V., Apaydin, A., Rodriguez, D. et al. (2015). RAL-1 controls multivesicular body biogenesis and exosome secretion. *J. Cell Biol.* 211 (1): 27–37.

87 Wei, Y., Wang, D., Jin, F. et al. (2017). Pyruvate kinase type M2 promotes tumour cell exosome release via phosphorylating synaptosome-associated protein 23. *Nat. Commun.* 8: 14041.

88 Verweij, F.J., Bebelman, M.P., Jimenez, C.R. et al. (2018). Quantifying exosome secretion from single cells reveals a modulatory role for GPCR signaling. *J. Cell Biol.* 217 (3): 1129–1142.

89 Ma, J., Zhang, S., Liu, J. et al. (2019). Targeted drug delivery to stroke via chemotactic recruitment of nanoparticles coated with membrane of engineered neural stem cells, Small. 15 (35): 1902011.

90 Leventis, P.A. and Grinstein, S. (2010). The distribution and function of phosphatidylserine in cellular membranes. *Annu. Rev. Biophys.* 39: 407–427.

91 Liu, F., Hu, S., Yang, H. et al. (2019). Hyaluronic acid hydrogel integrated with mesenchymal stem cell-secretome to treat endometrial injury in a rat model of Asherman's syndrome. *Adv. Healthc. Mater.* 8 (14): 1900411.

92 Muralidharan-Chari, V., Clancy, J., Plou, C. et al. (2009). ARF6-regulated shedding of tumor cell-derived plasma membrane microvesicles. *Curr. Biol.* 19 (22): 1875–1885.

93 Mathieu, M., Martin-Jaular, L., Lavieu, G., and Thery, C. (2019). Specificities of secretion and uptake of exosomes and other extracellular vesicles for cell-to-cell communication. *Nat. Cell Biol.* 21 (1): 9–17.

94 Tang, J., Vandergriff, A., Wang, Z. et al. (2017). A regenerative cardiac patch formed by spray painting of biomaterials onto the heart. *Tissue Eng. Part C* 23 (3): 146–155.

95 Nabhan, J.F., Hu, R., Oh, R.S. et al. (2012). Formation and release of arrestin domain-containing protein 1-mediated microvesicles (ARMMs) at plasma membrane by recruitment of TSG101 protein. *Proc. Natl. Acad. Sci. U.S.A.* 109 (11): 4146–4151.

96 Bianco, F., Perrotta, C., Novellino, L. et al. (2009). Acid sphingomyelinase activity triggers microparticle release from glial cells. *EMBO J.* 28 (8): 1043–1054.

97 Awojoodu, A.O., Keegan, P.M., Lane, A.R. et al. (2014). Acid sphingomyelinase is activated in sickle cell erythrocytes and contributes to inflammatory microparticle generation in SCD. *Blood* 124 (12): 1941–1950.

98 Del Conde, I., Shrimpton, C.N., Thiagarajan, P., and Lopez, J.A. (2005). Tissue-factor-bearing microvesicles arise from lipid rafts and fuse with activated platelets to initiate coagulation. *Blood* 106 (5): 1604–1611.

99 Lee, H., Li, C., Zhang, Y. et al. (2019). Caveolin-1 selectively regulates microRNA sorting into microvesicles after noxious stimuli. *J. Exp. Med.* 216 (9): 2202–2220.

100 Li, B., Antonyak, M.A., Zhang, J., and Cerione, R.A. (2012). RhoA triggers a specific signaling pathway that generates transforming microvesicles in cancer cells. *Oncogene* 31 (45): 4740–4749.

101 Schlienger, S., Campbell, S., and Claing, A. (2014). ARF1 regulates the Rho/MLC pathway to control EGF-dependent breast cancer cell invasion. *Mol. Biol. Cell* 25 (1): 17–29.

102 Bucki, R., Bachelot-Loza, C., Zachowski, A. et al. (1998). Calcium induces phospholipid redistribution and microvesicle release in human erythrocyte membranes by independent pathways. *Biochemistry* 37 (44): 15383–15391.

103 Pasquet, J.M., Dachary-Prigent, J., and Nurden, A.T. (1996). Calcium induces phospholipid redistribution and microvesicle release in human erythrocyte membranes by independent pathways. *Biochemistry* 37 (44): 15383–15391.

104 Wang, T., Gilkes, D.M., Takano, N. et al. (2014). Hypoxia-inducible factors and RAB22A mediate formation of microvesicles that stimulate breast cancer invasion and metastasis. *Proc. Natl. Acad. Sci.* 111 (31): E3234–E3242.

105 Kerr, J.F., Wyllie, A.H., and Currie, A.R. (1972). Apoptosis: a basic biological phenomenon with wideranging implications in tissue kinetics. *Br. J. Cancer* 26 (4): 239–257.

106 Elmore, S. (2007). Apoptosis: a review of programmed cell death. *Toxicol. Pathol.* 35 (4): 495–516.

107 Coleman, M.L., Sahai, E.A., Yeo, M. et al. (2001). Membrane blebbing during apoptosis results from caspase-mediated activation of ROCK I. *Nat. Cell Biol.* 3 (4): 339–345.

108 Sebbagh, M., Renvoizé, C., Hamelin, J. et al. (2001). Caspase-3-mediated cleavage of ROCK I induces MLC phosphorylation and apoptotic membrane blebbing. *Nat. Cell Biol.* 3 (4): 346–352.

109 Tixeira, R., Phan, T.K., Caruso, S. et al. (2020). ROCK1 but not LIMK1 or PAK2 is a key regulator of apoptotic membrane blebbing and cell disassembly. *Cell Death Differ.* 27 (1): 102–116.

110 Lee, N., MacDonald, H., Reinhard, C. et al. (1997). Activation of hPAK65 by caspase cleavage induces some of the morphological and biochemical changes of apoptosis. *Proc. Natl. Acad. Sci.* 94 (25): 13642–13647.

111 Rudel, T. and Bokoch, G.M. (1997). Membrane and morphological changes in apoptotic cells regulated by caspase-mediated activation of PAK2. *Science* 276 (5318): 1571–1574.

112 Tomiyoshi, G., Horita, Y., Nishita, M. et al. (2004). Caspase-mediated cleavage and activation of LIM-kinase 1 and its role in apoptotic membrane blebbing. *Genes Cells* 9 (6): 591–600.

113 Phan, T.K., Fonseka, P., Tixeira, R. et al. (2021). Pannexin-1 channel regulates nuclear content packaging into apoptotic bodies and their size. *Proteomics* 21 (13–14): e2000097.

114 Erwig, L.P. and Henson, P.M. (2008). Clearance of apoptotic cells by phagocytes. *Cell Death Differ.* 15 (2): 243–250.

115 Taylor, R.C., Cullen, S.P., and Martin, S.J. (2008). Apoptosis: controlled demolition at the cellular level. *Nat. Rev. Mol. Cell Biol.* 9 (3): 231–241.

116 Caruso, S. and Poon, I.K.H. (2018). Apoptotic Cell-Derived Extracellular Vesicles: More Than Just Debris. *Front. Immunol.* 9: 1486.

117 Rysavy, N.M., Shimoda, L.M., Dixon, A.M. et al. (2014). Beyond apoptosis: the mechanism and function of phosphatidylserine asymmetry in the membrane of activating mast cells. *Bioarchitecture* 4 (4–5): 127–137.

118 Lunavat, T.R., Cheng, L., Kim, D.K. et al. (2015). Small RNA deep sequencing discriminates subsets of extracellular vesicles released by melanoma cells – Evidence of unique microRNA cargos. *RNA Biol.* 12 (8): 810–823.

119 Di Vizio, D., Kim, J., Hager, M.H. et al. (2009). Oncosome formation in prostate cancer: association with a region of frequent chromosomal deletion in metastatic disease. *Cancer Res.* 69 (13): 5601–5609.

120 Valcz, G., Buzas, E.I., Szallasi, Z. et al. (2018). Perspective: bidirectional exosomal transport between cancer stem cells and their fibroblast-rich microenvironment during metastasis formation. *NPJ Breast Cancer* 4: 18.

121 Terrasi, A., Bertolini, I., Martelli, C. et al. (2019). Specific V-ATPase expression sub-classifies IDHwt lower-grade gliomas and impacts glioma growth in vivo. *EBioMedicine* 41: 214–224.

122 Kim, J., Morley, S., Le, M. et al. (2014). Enhanced shedding of extracellular vesicles from amoeboid prostate cancer cells: potential effects on the tumor microenvironment. *Cancer Biol. Ther.* 15 (4): 409–418.

123 Hager, M.H., Morley, S., Bielenberg, D.R. et al. (2012). DIAPH3 governs the cellular transition to the amoeboid tumour phenotype. *EMBO Mol. Med.* 4 (8): 743–760.

124 De Silva, R.T., Dissanayake, R.K., Mantilaka, M. et al. (2018). Drug-loaded halloysite nanotube-reinforced electrospun alginate-based nanofibrous scaffolds with sustained antimicrobial protection. *ACS Appl. Mater. Interfaces* 10 (40): 33913–33922.

125 Pournaghi, M., Khodavirdilou, R., Saadatlou, M.A.E. et al. (2021). Effect of melatonin on exosomal dynamics in bovine cumulus cells. *Process Biochem.*

126 Chukhchin, D.G., Bolotova, K., Sinelnikov, I. et al. (2019). Exosomes in the phloem and xylem of woody plants. *Planta.*

127 Vijayarathna, S., Chen, Y., Kanwar, J.R., and Sasidharan, S. (2017). Standardized Polyalthia longifolia leaf extract (PLME) inhibits cell proliferation and promotes apoptosis: The anti-cancer study with various microscopy methods. *Biomed. Pharmacother.*

128 Beekman, P., Enciso-Martinez, A., Rho, H.S. et al. (2019). Immuno-capture of extracellular vesicles for individual multi-modal characterization using AFM, SEM and Raman spectroscopy. *Lab Chip* 19 (15): 2526–2536.

129 Bagrov, D.V., Senkovenko, A.M., Nikishin, I.I. et al. (2021). Application of AFM, TEM, and NTA for characterization of exosomes produced by placenta-derived mesenchymal cells. *J. Phys. Conf. Ser.* 1942 (1): 012013.

130 Wang, L., Yang, L., Zhuang, T., and Shi, X. (2022). Tumor-derived exosomal miR-29b reduces angiogenesis in pancreatic cancer by silencing ROBO1 and SRGAP2. *J. Immunol. Res.* 2022: 4769385.

131 Yuan, Y., Zhang, Q., Yan, Y. et al. (2018). Designed construction of tween 60@2β-CD self-assembly vesicles as drug delivery carrier for cancer chemotherapy. *Drug Delivery* 25 (1): 623–631.

132 Sharma, S., LeClaire, M., and Gimzewski, J.K. (2018). Ascent of atomic force microscopy as a nanoanalytical tool for exosomes and other extracellular vesicles. *Nanotechnology* 29 (13): 132001.

133 Sharma, S., Rasool, H.I., Palanisamy, V. et al. (2010). Structural-mechanical characterization of nanoparticle exosomes in human saliva, using correlative AFM, FESEM, and force spectroscopy. *ACS Nano* 4 (4): 1921–1926.

134 Ali, A.G., Mohamed, M.F., Abdelhamid, A.O., and Mohamed, M.S. (2016). A novel adamantane thiadiazole derivative induces mitochondria-mediated apoptosis in lung carcinoma cell line. *Bioorg. Med. Chem.* 25 (1): 241–253.

135 Sharma, S., LeClaire, M., Wohlschlegel, J., and Gimzewski, J. (2020). Impact of isolation methods on the biophysical heterogeneity of single extracellular vesicles. *Sci. Rep.* 10 (1): 13327.

136 Beckers, M., Mann, D., and Sachse, C. (2020). Structural interpretation of cryo-EM image reconstructions. *Prog. Biophys. Mol. Biol.* 160: 26–36.

137 Bairamukov, V.Y., Bukatin, A.S., Kamyshinsky, R.A. et al. (2022). Nanomechanical characterization of exosomes and concomitant nanoparticles from blood plasma by PeakForce AFM in liquid. *Biochim. Biophys. Acta* 1866 (7): 130139.

138 Matatyaho Ya'akobi, A. and Talmon, Y. (2021). Extending cryo-EM to nonaqueous liquid systems. *Acc. Chem. Res.* 54 (9): 2100–2109.

139 Jiang, X., Sun, J., Zuckermann, R.N. et al. (2017). Accurate cryo-EM characterizations of polypeptoid vesicles. *Microsc. Microanal.* 23 (S1): 836–837.

140 McNicholas, K. and Michael, M.Z. (2016). Immuno-characterization of exosomes using nanoparticle tracking analysis. *Methods Mol. Biol.* 1545: 35–42.

141 Gouin, K., Perk, K., Antes, T. et al. (2017). A comprehensive method for identification of suitable reference genes in extracellular vesicles. *J. Extracell. Vesicles* 6 (1): 1347019.

142 Fan, M., Liu, H., Yan, H. et al. (2022) A CAR T-inspiring platform based on antibody-engineered exosomes from antigen-feeding dendritic cells for precise solid tumor therapy. *Biomaterials* 282: 121424.

143 Liu, Y., Wu, S., Koo, Y. et al. (2020). Characterization of and isolation methods for plant leaf nanovesicles and small extracellular vesicles. *Nanomedicine* 29: 102271.

144 Lee, T. S., Kim, Y., Zhang, W. et al. (2018). Facile metabolic glycan labeling strategy for exosome tracking. *Biochim. Biophys. Acta Gen. Subj.* 1862 (5): 1091–1100.

145 Tae Seong, L., Yoojin, A., Young-Jun, I. et al. (2021). The characterization of exosomes from fibrosarcoma cell and the useful usage of dynamic light scattering (DLS) for their evaluation. *PLoS One* 16 (1): e0231994.

146 Hosseinzadeh, S., Nazari, H., Esmaeili, E. et al. (2021). Polyethylene glycol triggers the anti-cancer impact of curcumin nanoparticles in sw-1736 thyroid cancer cells. *J. Mater. Sci. - Mater. Med.* 32 (9): 112.

147 Sarra, A., Celluzzi, A., Bruno, S.P. et al. (2020). Biophysical characterization of membrane phase transition profiles for the discrimination of outer membrane vesicles (OMVs) from escherichia coli grown at different temperatures. *Front. Microbiol.* 11: 290.

148 Choi, D., Montermini, L., Jeong, H. et al. (2019). Mapping subpopulations of cancer cell-derived extracellular vesicles and particles by nano-flow cytometry. *ACS Nano* 13 (9): 10499–10511.

149 Que, Z.-J., Luo, B., Wang, C.-T. et al. (2020). Proteomics analysis of tumor exosomes reveals vital pathways of Jinfukang inhibiting circulating tumor cells metastasis in lung cancer. *J. Ethnopharmacol.* 256: 112802.

150 Wang, X.-Y., Yang, H., Wang, M.-G. et al. (2017). Trehalose protects against cadmium-induced cytotoxicity in primary rat proximal tubular cells via inhibiting apoptosis and restoring autophagic flux. *Cell Death Dis.* 8 (10): e3099.

151 Domenis, R., Zanutel, R., Caponnetto, F. et al. (2017). Characterization of the proinflammatory profile of synovial fluid-derived exosomes of patients with osteoarthritis. *Mediators Inflammation* 2017: 4814987.

152 Görgens, A. and Nolan, J.P. (2020). Aiming to compare apples to apples: analysis of extracellular vesicles and other nanosized particles by flow cytometry. *Cytometry, Part A* 97 (6): 566–568.

153 Busatto, S., Giacomini, A., Montis, C. et al. (2018). Uptake profiles of human serum exosomes by murine and human tumor cells through combined use of colloidal nanoplasmonics and flow cytofluorimetric analysis. *Anal. Chem.* 90 (13): 7855–7861.

154 Wang, Y., Rong, J., Zhang, J. et al. (2016). Morphology, in vivo distribution and antitumor activity of bexarotene nanocrystals in lung cancer. *Drug Dev. Ind. Pharm.* 43 (1): 132–141.

155 Morales-Kastresana, A., Telford, B., Musich, T.A. et al. (2017). Labeling extracellular vesicles for nanoscale flow cytometry. *Sci. Rep.* 7 (1): –1878.

156 Liu, C., Xu, X., Li, B. et al. (2018). Single-exosome-counting immunoassays for cancer diagnostics. *Nano Lett.* 18 (7): 4226–4232.

157 Jia, R., Rotenberg, S.A., and Mirkin, M.V. (2022). Electrochemical resistive-pulse sensing of extracellular vesicles. *Anal. Chem.* 94 (37): 12614–12620.

158 Shao, H., Im, H., Castro, C.M. et al. (2018). New technologies for analysis of extracellular vesicles. *Chem. Rev.* 118 (4): 1917–1950.

159 Im, H., Shao, H., Park, Y.I. et al. (2014). Label-free detection and molecular profiling of exosomes with a nano-plasmonic sensor. *Nat. Biotechnol.* 32 (5): 490–495.

3

Therapeutic Potential of Extracellular Vesicles from Different Cell Sources

Xueyi Wang[1,2] *and Zhenhua Li*[1,2]

[1] *The Tenth Affiliated Hospital of Southern Medical University, 78 Wanjiang Avenue, Dongguan, Guangdong 523059, China*
[2] *Guangdong Provincial Key Laboratory of Shock and Microcirculation, Southern Medical University, Shatai South Avenue Guangzhou, Guangdong 510080, China*

Extracellular vesicles (EVs) carry a wide range of bioactive molecules, such as proteins, lipids, and nucleic acids, and have emerged as important mediators of intercellular communication. In recent years, the therapeutic potential of EVs has garnered significant attention as a potential treatment and as a biomarker for various pathologies. Accordingly, studies have shown that EVs originating from diverse cell types play different roles in biological phenomena, due to the cargo including a common set of proteins of EVs from different cell types reflecting the cell source of the EVs, and the pathological and physiological state of the cell source [1]. Thus, EVs from stem cells, including mesenchymal stem cells (MSCs), neural stem cells (NSCs), endothelial progenitor cells (EPCs), and cardiac progenitor cells (CPCs), immune cells, such as NK cells, macrophage, dendritic cells (DCs), T cells, cancer cells, and plant sources (broccoli, ginger, garlic, carrot, grape, blueberry, strawberry, apple and lemon), have been widely demonstrated for applications in organ transplantation, human immunodeficiency virus (HIV), cardiovascular diseases, neuroprotection, cancer therapy, and regenerative medicine.

In this chapter, we will provide an overview of the therapeutic potential of EVs from different cell sources.

3.1 Extracellular Vesicles Derived from Stem Cells (SCs)

Stem cells (SCs) are cells that have the potential to differentiate into various cell types, and they have been extensively studied for their therapeutic potential in a variety of diseases. Stem cells and progenitor cells have emerged as the most promising therapeutic options for the treatment of degenerative or genetic disorders that were previously considered incurable. Hence, the emergence of MSCs, NSCs, EPCs, and CPCs has raised hopes of treating nonhematopoietic ailments through

Biomedical Applications of Extracellular Vesicles, First Edition.
Edited by Zhenhua Li, Xing-Jie Liang, and Ke Cheng.

the replacement of damaged cells with new ones derived from these stem cells. MSCs are multipotent stromal cells, known for their ability to differentiate into various mesodermal cell types, including adipocytes, chondrocytes, and osteoblasts [2]. Due to their multipotency, easy isolation from adult tissues, and large ex vivo expansion capacity, MSCs are widely used in clinical trials [3]. EPCs are a subgroup of bone marrow-derived cells that share similar cell surface markers with vascular endothelial cells. They adhere to endothelium in areas with hypoxia or ischemia, and contribute to the formation of new blood vessels [4, 5]. CPCs are CPCs found in the heart that may originate from either the embryonic cell population or bone marrow. These cells are thought to participate in the normal turnover of cardiac myocytes and vascular endothelial cells [6, 7]. NSCs are multipotent, self-renewing cells that can be obtained from the fetal and adult brain with the potential to differentiate into oligodendrocytes, astrocytes, and neurons [8].

Increasing evidence indicating that EVs play a crucial role in intercellular communication in various cell types (Figure 3.1) [9, 10] suggests that EVs may also serve a similar purpose in facilitating communication between stem cells and other cells. Therefore, it is reasonable to assume that EVs could be essential in the theory that stem cells exert therapeutic effects by releasing signals that communicate with recipient cells, triggering repair and regeneration processes.

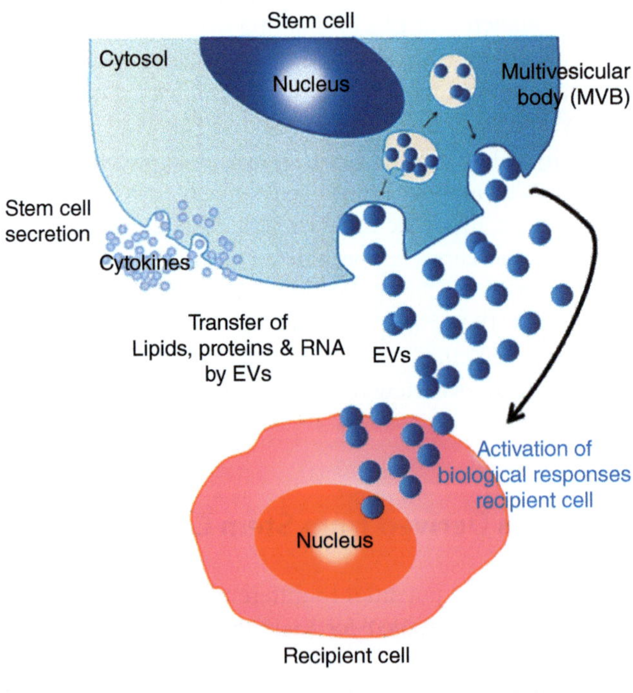

Figure 3.1 A proposed model for mechanisms underlying stem cell EVs therapeutic potential. EVs transfer lipids, proteins, and genetic material to recipient cells, leading to their biological activation and the desired therapeutic effect. Source: Zhang et al. [9]/MDPI/CC BY 4.0.

3.1.1 Extracellular Vesicles Derived from Mesenchymal Stem Cells (MSCs)

Initially, MSCs caught the attention of researchers due to their capacity to differentiate into mesodermal cell lineages both *in vitro* and *in vivo* [11, 12]. However, recent studies have shown that the therapeutic effects of MSCs are not primarily due to their differentiation ability [13]. In fact, MSCs possess other critical properties, including transdifferentiation into non-mesodermal lineages and their ability to migrate to the site of injury, interact with host cells, and release various paracrine factors and growth factors. These factors can modulate the immune response, alter the responses of the endothelium and epithelium to injury, and promote tissue regeneration (Figure 3.2).

Over the past few years, numerous research teams have explored the therapeutic potential of MSCs-derived EVs. The efficacy of these EVs has been observed in various diseases such as kidney injury, myocardial ischemia/reperfusion injury (MI/RI), spinal cord injury (SCI), and cancer. Despite limited reports, these findings indicated that MSCs-derived EVs mimic the characteristics of their parent MSCs and exhibit immense therapeutic potential in treating various diseases.

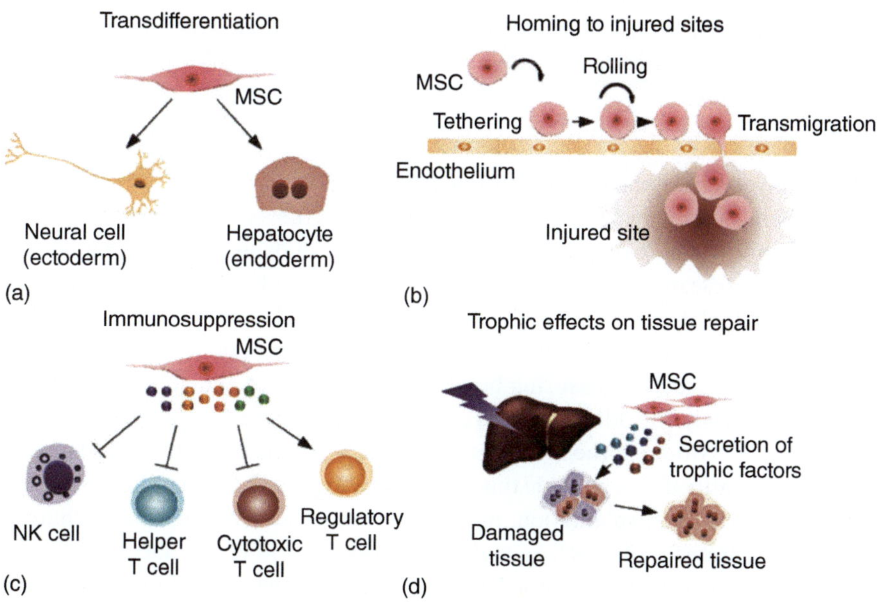

Figure 3.2 Various therapeutic effects of MSCs. Source: Katsuda et al. [13]/John Wiley & Sons. (a) MSCs have the ability to differentiate into cells that do not originate from mesoderm, such as hepatocytes and neurons. (b) MSCs can migrate to injured tissues, although the exact mechanism of how they move across endothelial barriers and target specific receptors or ligands produced by damaged tissues is not fully understood. This process resembles the recruitment of leukocytes during inflammation. (c) MSCs can suppress the immune response by releasing cytokines that inhibit natural killer cells, helper T cells, and cytotoxic T cells, while also promoting the generation of regulatory T cells. (d) MMSCs can produce trophic factors that facilitate the regeneration of damaged tissue.

3.1.1.1 Kidney Injury

The initial investigation into the therapeutic potential of MSCs-derived EVs was focused on their ability to treat kidney damage [14]. It is widely acknowledged that MSCs contribute to the restoration of injured nephrons by temporarily migrating to the renal vasculature, rather than directly incorporating into regenerating tubules. This suggested that MSCs supported the intrinsic repair mechanisms employed by surviving epithelial cells in a paracrine manner. In fact, MSCs have been demonstrated to safeguard the kidney from toxic injury by producing factors that reduce apoptosis and promote the proliferation of endogenous tubular cells. As part of this paracrine effect, Camussi's group confirmed that EVs derived from bone marrow-derived MSCs could protect against acute kidney injury (AKI) in a mouse model induced by glycerol [15]. The regenerative effects of EVs and MSCs on the recovery of AKI were found to be similar, as both induced proliferation and reduced apoptosis in tubular epithelial cells. This was further demonstrated in another study that also used a severe form of AKI induced by cisplatin [16]. Additionally, the study showed that a single administration of MSCs-derived EVs immediately after MI/RI protected against the development of both acute and chronic kidney injury [17]. The researchers discovered that EVs transport a specific subset of cellular mRNAs, including those associated with mesenchymal features, transcription control, proliferation, and immunoregulation, thereby emphasizing the potential of MSCs-derived EVs for regenerative medicine [15, 16].

3.1.1.2 Myocardial Ischemia/Reperfusion Injury (MI/RI)

MSCs' therapeutic impact on MI/RI is primarily achieved by either minimizing tissue damage or improving tissue regeneration. In a study using pigs as a model for MI/RI, Lim's research team demonstrated that treatment with MSCs-conditioned medium resulted in a decrease in the size of the myocardial infarction [18]. Furthermore, the report revealed that the active component in this fraction was found to be enriched with particles measuring between 50 and 200 nm, which were identified as EVs. Conversely, the other fractions did not demonstrate any protective effects on MI/RI [19, 20]. Although the molecules responsible for the protective effects of EVs are not yet identified, it is possible that cytokines and growth factors, similar to those involved in MSCs transplantation, may play a role. Another potential mechanism could be the transfer of miRNAs and/or mRNAs from MSCs to damaged cardiac cells, which may have therapeutic benefits. Moreover, the clinical use of naturally derived EVs is limited by their poor ability to effectively target and home to injured myocardium. To address this challenge, Zhang et al. developed EV-monocyte mimics (Mon-Exos) via membrane fusion technology (Figure 3.3) [21]. After MI, monocytes as inflammatory cells are extensively recruited to the damaged heart and undergo differentiation into macrophages at the site of injury. Mon-Exos combined the regenerative features of stem cells with the targeting capabilities of monocytes, enabling them to effectively locate and repair the affected region. In a model of MI/RI, the Mon-Exos demonstrated the ability to promote angiogenesis and

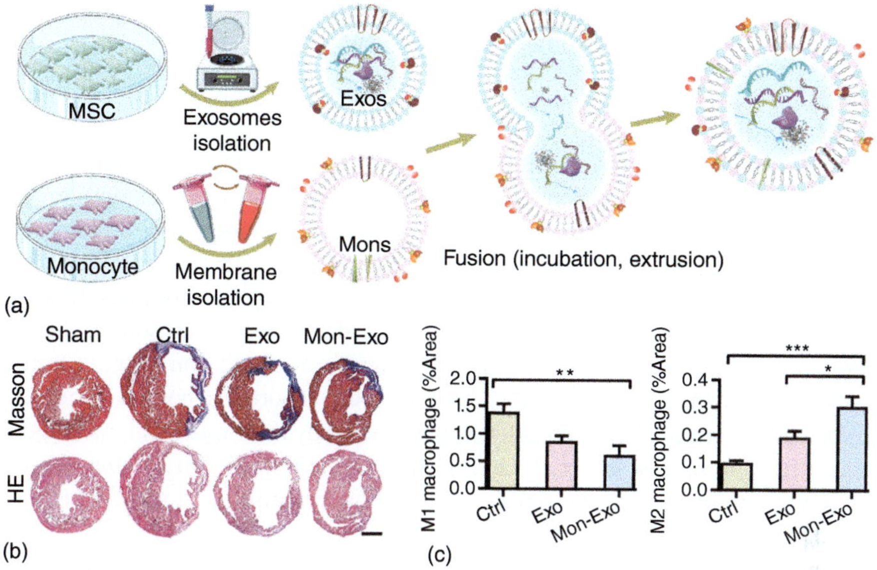

Figure 3.3 The therapeutic effect of MSCs-derived EVs on MI/RI. Source: Zhang et al. [21]]/with permission of Elsevier. (a) The process of constructing EV-Monocyte mimics (Mon-Exos) can be represented by a schematic diagram. (b) After four weeks of treatment, heart sections from a mouse model of MI/RI were analyzed through Masson and HE staining to evaluate the volumes of collagen and infarct areas. (c) Mon-Exos were found to effectively promote M2 polarization and decrease the population of M1 cells, thereby regulating the homeostasis of macrophage subpopulations.

modulate macrophage subpopulations, thereby regulating the progression of the disease.

3.1.1.3 Spinal Cord Injury (SCI)

Researchers have utilized the anti-inflammatory properties of MSCs-derived EVs to promote spinal cord repair by modulating the inflammatory environment and promoting regeneration. In a recent study, the researchers incorporated MSCs-derived EVs into a conductive hydrogel to address the issue of exacerbated inflammation caused by the implantation of repair materials in mice. The synergistic treatment of encapsulating MSCs-derived EVs in the conductive hydrogel significantly inhibited early inflammation followed by SCI. Furthermore, the hydrogel's binding with EVs resulted in a gradual and prolonged release of EVs during the initial stages of implantation. Furthermore, a more manageable and regulated delivery platform for EVs was built to eliminate the risk of the secondary injuries and limited efficacy. A recent study reported a microneedle array patch for local implantation using gelatin methacryloyl (GelMA) mixed with EVs derived from MSCs that were cultured in a three-dimensional (3D) environment (GelMA-MN@3D-Exo). Compared to traditional two-dimensional (2D) culture, MSCs cultured in a 3D environment maintain their stem cell properties, resulting in a significant enhancement in the therapeutic effectiveness of their secreted EVs (3D-Exo). These secreted EVs contain more active proteins and miRNAs that are involved in regulating the local microenvironment, thus providing better therapeutic benefits on SCI.

3.1.1.4 Cancer

Recently, there have been some studies that reported the potential antitumor properties of exosomes derived from MSCs. Lee et al. reported that exosomes derived from MSCs can impede tumor growth and hinder the formation of new blood vessels by decreasing the expression of vascular endothelial growth factor (VEGF) [22]. This effect was achieved through the delivery of miR-16 by the exosomes. In addition, Katakowski et al. discovered that modifying the content of exosomes, such as increasing the expression of miR-146, led to significant inhibition of brain tumor growth [23]. Furthermore, bone MSCs have also been reported to possess some antitumor activity, but further application was restricted by their weak and insufficient effects. To address this, Ma et al. developed a new approach that combined bone MSCs with tumor-derived exosomes, and the enhanced antitumor ability was confirmed [24].

3.1.2 Extracellular Vesicles Derived from Neural Stem Cells (NSCs)

NSCs possess the capacity to self-renew and differentiate into various cell types of the central nervous system, such as neurons, astrocytes, and oligodendrocytes [25, 26]. Thus, NSCs have been chosen as valuable tools for investigating the underlying mechanisms of various central nervous system diseases due to their unique characteristics.

The use of EVs derived from NSCs has been proposed as a potential strategy to address oxidative stress and acute neuroinflammation following traumatic brain injury (TBI) [27]. This is due to the presence of antioxidants and anti-inflammatory substances within the EVs. Recent researches has revealed that EVs can be utilized as a carrier to transport small molecules and drugs to the brain [28–30]. This targeted delivery system can be employed to deliver antioxidants to neurons in order to reduce the levels of reactive oxygen species (ROS). This promising approach has the potential to treat various neurological conditions and diseases such as Alzheimer's and Parkinson's disease [31–33]. Moreover, Sun et al. investigated a study to assess the ability of human NSCs-EVs to offer neuroprotection in a controlled cortical impact injury (CCI) model (Figure 3.4) [34]. The NSCs-EVs were extracted from human embryonic H9 cell culture media using ultrafiltration method. The investigation demonstrated that the administration of NSCs-EVs increased the movement of naturally occurring NSCs towards the injury site in both male and female rats. In male rats, NSCs-EVs also resulted in the expression of vascular endothelial growth factor receptor 2 (VEGFR2) without affecting the total vascular density. The study's outcomes indicated that NSC-EV therapy may improve neuroprotection and restore motor function in male rats after TBI, accompanied by enhanced migration of endogenous NSCs towards the lesion site and heightened VEGF activity.

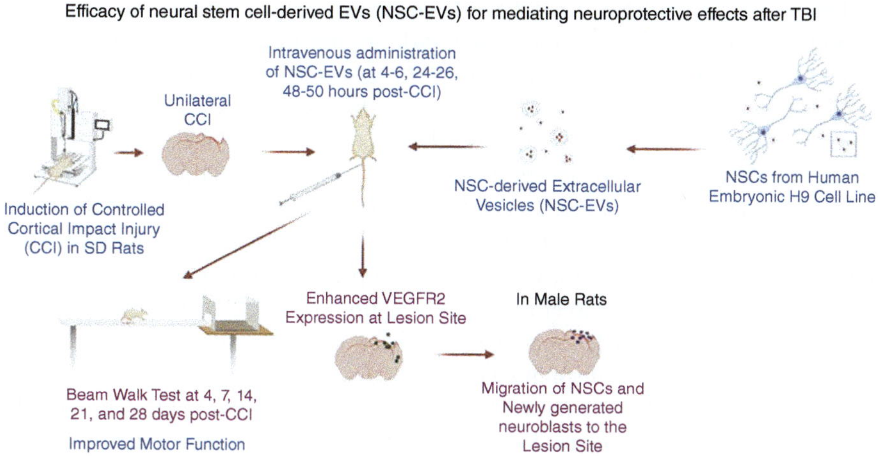

Figure 3.4 Efficacy of neural stem cell-derived EVs in animal models of traumatic TBI. Source: Hering and Shetty [27]/Oxford university Press/CC BY 4.0.

3.1.3 Extracellular Vesicles Derived from Endothelial Progenitor Cells (EPCs)

EPCs possess the ability to transform into endothelial cells, which make up the interior lining of blood vessels [35]. Through their exosomes, EPCs can influence angiogenesis—a process essential for the growth and development of new blood vessels. The exosomes derived from EPCs attach to specific integrins (namely, α4 and β1 integrins) expressed on the surface of microvesicles, promoting endothelial cell survival, proliferation, and organization both *in vitro* and *in vivo* [36]. Recent studies have highlighted the potential of EVs derived from EPCs as a therapeutic agent in various diseases. These EVs are involved in angiogenesis and antiapoptotic processes, and their contents, such as microRNAs, determine their effects. In particular, EVs containing miR-126 and miR-296 have been found to promote angiogenesis and antiapoptosis in rat models of MI/RI and murine models of hind limb ischemia [37, 38]. EVs released from EPCs have also been shown to protect human islets by enhancing their vascularization as well as protect against acute kidney injury [39]. Furthermore, exosomes derived from EPCs have demonstrated protective effects on cardiomyocytes against Ang II-induced hypertrophy and apoptosis [40]. In various disease models, EVs have yielded distinct effects based on the specific RNAs they carried. Overall, these findings suggest that EVs derived from EPCs hold promise as a therapeutic agent.

3.1.4 Extracellular Vesicles Derived from Cardiac Progenitor Cells (CPCs) and Other Stem Cells

CPCs are a promising candidate for treating myocardial diseases due to their ability to differentiate into cardiovascular lineages and produce paracrine factors [41, 42]. The paracrine effects of CPCs are primarily mediated through exosomes, which contain microRNAs that have been shown to promote cardio protection in both mouse and human models [43, 44]. Studies by Gray et al. have demonstrated that hypoxic conditions induce CPCs to secrete proregenerative exosomes with increased levels of miRNAs compared to normal exosomes [45]. These hypoxia-induced exosomes have been shown to improve cardiac function and reduce fibrosis *in vivo*. In addition, Ong et al. observed that overexpression of hypoxia-inducible factor-1 (HIF-1) in CPCs led to improved survival of the transplanted cells [46]. This was attributed to the presence of high levels of miR-126 and miR-120 in the exosomes, which activated prosurvival kinases and induced a glycolytic switch in the recipient CPCs. These findings suggested that transferring microRNAs from host cells to transplanted cells may be a promising approach to enhance the survival of transplanted cells.

EVs derived from other stem cell types have also been found to play crucial roles in various biological processes. For example, EVs obtained from human liver stem cells (HLSCs) have been confirmed to promote self-renewal and expansion of stem cells [47]. Moreover, exosomes derived from HLSCs are capable of activating a proliferative program in remnant hepatocytes following hepatectomy by transferring specific mRNAs horizontally, which ultimately accelerates liver regeneration *in vivo* [48]. Furthermore, exosomes secreted by human CD34(+) stem cells have exhibited independent angiogenic activity both *in vitro* and *in vivo* [49]. These findings indicated that EVs derived from stem cells were significant contributors to the paracrine effect of progenitor cell transplantation, particularly for therapeutic angiogenesis.

3.2 Extracellular Vesicles Derived from Immune Cells

The immune system is a highly complex network consisting of numerous cell types that are distributed throughout different organs in the body. It serves the dual purpose of maintaining cellular equilibrium and defending the host against foreign invaders. To achieve these goals, the immune system has evolved various communication pathways, including direct cell-to-cell interactions and the secretion of soluble molecules. Another mode of information exchange utilized by immune cells is the release of extracellular vesicles, primarily exosomes, which have been extensively investigated [50].

EVs derived from immune cells have the ability to elicit either immune-stimulatory or immune-inhibitory responses, depending on the parental cell type and its state. These EVs can thus serve dual functions in physiological and pathological processes. There is abundant evidence demonstrating that EVs obtained from various immune cells, such as macrophages, dendritic cells (DCs), T cells, and natural killer (NK) cells, possess distinct functionalities (Figure 3.5) [51].

3.2.1 Extracellular Vesicles Derived from Macrophages

Tumor-associated macrophages (TAMs) can be categorized into two types: M1 macrophages, which promote an antitumor microenvironment, and M2

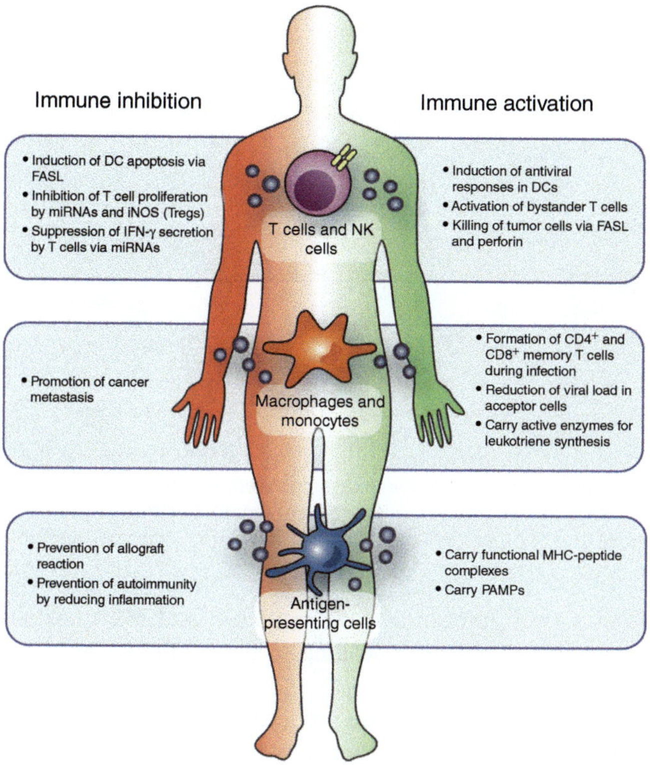

Figure 3.5 EVs derived from immune cells play distinct roles in regulating immune responses. The functions of these EVs are dependent on the specific properties of their parent cells. For instance, immune cell-derived EVs can stimulate immune responses against viral infections and tumors by activating DCs and T cells through various mechanisms. On the other hand, they can also inhibit DCs and T cells, which can result in cancer metastasis, prevention of allograft reaction, and autoimmunity. These EVs contain different components such as FAS ligand (FASL), interferon-g (IFN-g), inducible nitric oxide synthase (iNOS), pathogen-associated molecular pattern (PAMP), natural killer cells (NK), and regulatory T cells (Treg). Source: Veerman et al. [50]/with permission of Elsevier.

macrophages, which promote a protumor microenvironment. M1 macrophages release proinflammatory cytokines, while M2 macrophages release anti-inflammatory cytokines. Notably, macrophages exhibit remarkable flexibility and can transition between these two phenotypes [52]. The EVs produced by M1 macrophages also possess proinflammatory capabilities inherited from their parent cells. EVs can be engineered to enhance the immune system for cancer treatment by effectively targeting and increasing the proinflammatory capacity [53]. Gunassekaran et al. confirmed this by modifying EVs derived from M1 macrophages to target TAMs and reprogram them into the M1 subtype [54]. To target the TAMs' interleukin-4 receptor (IL4R), the researchers conjugated an IL4R-binding peptide onto the surface of EVs. To promote polarization towards the M1 subtype, they transfected M1 type EVs with NF-κB p50 siRNA and miR-511-3p, resulting in IL4R-Exo(si/mi). When IL4R-Exo (si/mi) was systemically administered, the target

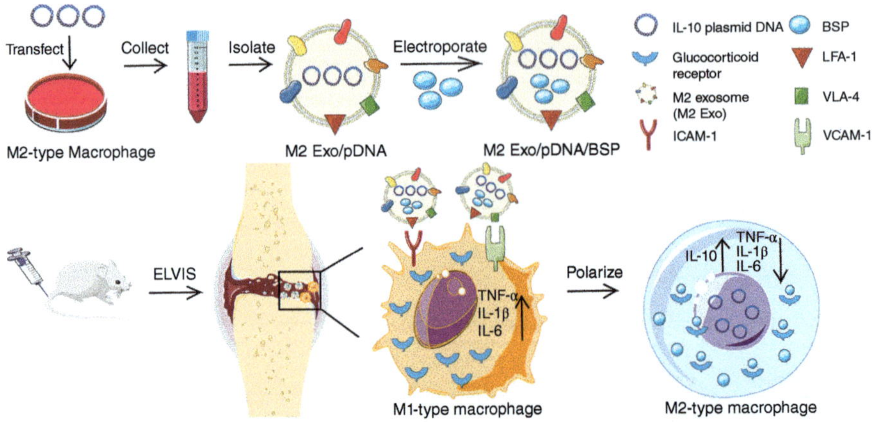

Figure 3.6 Exosomes derived from M2 macrophages target areas of inflammation while also possessing anti-inflammatory properties. These exosomes were engineered to carry both IL-10 pDNA and GCs, which can work together to treat rheumatoid arthritis by repolarizing M1 macrophages into M2 macrophages. Source: Li et al. [55]/with permission of Elsevier.

genes in M2 macrophages were significantly downregulated, and M2 cytokines were reduced while M1 cytokines increased, leading to the inhibition of tumor growth.

M2 macrophage-derived EVs possess the same properties as M2 macrophages, including homing capabilities and anti-inflammatory functions. Utilizing these EVs as a delivery system for substances like miRNAs and IL-10 can effectively enhance the polarization of macrophages towards the M2 type, resulting in improved treatment efficacy (Figure 3.6) [55, 56]. Tang et al. reported a novel approach to produce IL-10 loaded EVs by engineered macrophages [57]. Furthermore, the therapeutic potential of these EVs in a mouse model of acute renal injury caused by ischemia was evaluated. The researchers found that the presence of integrins on the surface of the EVs facilitated their stability and ability to target the kidney. Moreover, the IL-10 loaded EVs efficiently targeted macrophages in the tubulointerstitial region and stimulated their polarization in the presence of IL-10. Treatment with these EVs resulted in a significant reduction in renal tubular cell damage and inflammation caused by MI/RI, ultimately leading to the prevention of chronic kidney disease.

3.2.2 Extracellular Vesicles Derived from Dendritic Cells (DCs)

In the recent years, researchers have extensively studied the potential of using EVs derived from DCs as a means of delivering antigens for anticancer treatments. Clinical trials have shown that using these EVs in cancer therapy is both feasible and safe, although immune responses have been inadequate [58, 59]. To address this issue, various modification strategies have been explored to enhance the immune-stimulatory properties of DC-derived EVs, resulting in stronger immune responses. For example, codelivery of tumor peptides with alpha-galactosylceramide has been shown to induce adaptive antitumor immunity

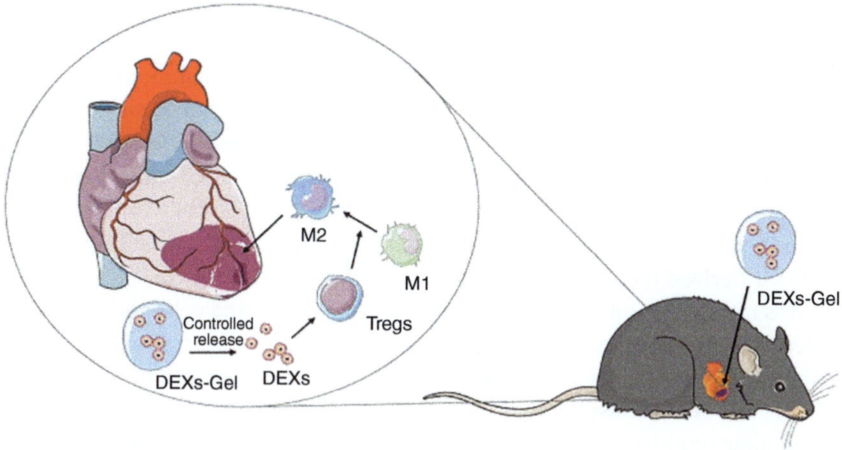

Figure 3.7 The schematic figure of DCs-derived exosomes therapeutic effects after MI by activating Treg cells and modifying the polarization of macrophages. Source: Zhang et al. [63]/Springer nature/CC BY 4.0.

in a mouse model of melanoma [60]. And EVs secreted by bone marrow DCs have been found to elicit improved antitumor activity upon Toll-like receptor 3 (TLR3) stimulation in DCs. In a mouse model of hepatocellular carcinoma, DC-derived EVs expressing tumor antigen alpha-fetoprotein and transfected with lentivirus were able to suppress tumor progression by reducing the number of Treg cells and increasing the percentage of $CD8^+$ T cells in the tumor microenvironment [61].

Additionally, it has been proposed that the therapeutic benefits of EVs derived from DCs are not restricted to treating cancer, and these EVs may be capable of triggering an immune response against HIV-1 infection through the controlled release of EVs at low doses. Ellwanger et al. confirmed that exosomes derived from DCs can serve as a precise immune regulatory tool, but their effectiveness may not be generalized, as a low release of exosomes could be more advantageous for the DCs-derived immune therapy (IT) response than a high release [62].

Besides, Zhang et al. reported the systemic delivery of EVs secreted from DCs could mediate Treg cell activation and improve wound healing post-MI in mice, providing a promising new therapeutic option available for the damaged myocardium caused by a myocardial infarction (Figure 3.7) [63].

3.2.3 Extracellular Vesicles Derived from T Cells

Compared to extensively studied DCs-derived EVs, less research has been done on T cell-derived EVs. Once T-cell receptor (TCR) is activated, T cells release exosomes containing the TCR/CD3 complex in high amounts [64]. Like their parental cells, T cell-derived exosomes have primarily been associated with antiviral and antitumor responses. For instance, T cells release exosomes containing mitochondrial DNA at an immunological synapse, which is transferred to DCs. Thus, T cell-derived exosomes provide a signal that can trigger an antiviral response *in vitro*, resulting in DCs less susceptible to viral infection [65].

In addition, $CD8^+$ T cell-derived EVs exhibit remarkable capabilities that could be harnessed for therapeutic purposes, particularly in cancer treatment. These EVs can trigger programmed cell death in MSCs within the tumor, effectively impeding tumor growth [66]. When used in combination with cell-based immunotherapy, $CD8^+$ T cell-derived EVs have the potential to enhance the efficacy of the therapy, as the tumor microenvironment often impedes their effects. Thus, by administering $CD8^+$ T cell-derived EVs, it is possible to disrupt the tumor microenvironment, creating a less immune-inhibitory environment for the transferred cells and improving the efficacy of immunotherapy.

Furthermore, Tregs are vital for maintaining immune balance, and EVs derived from Tregs are being considered as a potential for transplantation therapy as they exhibit similar immunomodulatory properties to their parent cells. These EVs can hinder cell-cycle progression and impede T cell proliferation, leading to enhanced rat survival in a kidney transplantation model by carrying specific miRNAs and iNOS. Additionally, Treg-derived EVs can prevent widespread inflammation by transporting IFN-g-suppressing miRNAs to T helper cells [67]. Furthermore, studies also suggest that their EVs can alter DC function by delivering cytokine-suppressive miRNAs. Overall, Treg-derived EVs may serve as immunomodulators to promote a more immunosuppressive environment.

3.2.4 Extracellular Vesicles Derived from Natural Killer (NK) Cells

NK cells play a crucial role in the immune system's defense against cancer and pathogens by destroying target cells through the release of cytotoxic molecules or by inducing apoptosis via ligation of activation receptors. NK cell-derived exosomes contain typical NK cell molecules such as CD56, NKG2D, FASL, perforin, and granzymes, which can induce apoptosis of tumor cell lines *in vitro* [50, 68–70]. This makes NK cell-derived EVs promising candidates for immunotherapy. Recent studies have demonstrated that NK cell-derived EVs exhibit potent antitumor effects *in vivo* through the actions of FASL and perforin, as observed in mouse xenograft models of glioblastoma and melanoma [71, 72]. Notably, in the glioblastoma study, dextran sulfate was administered to mice to inhibit scavenger receptor A, thereby preventing EVs clearance by the liver, as consistent with the previous reports [73]. By inhibiting EVs clearance in the liver, the EVs were found to accumulate significantly in the spleen and tumor, suggesting that this strategy could be useful for improving the targeting of EVs-based therapeutics.

Unlike T cells, NK cells do not require MHC class I to recognize target cells. Tumor and virus-infected cells often downregulate MHC class I to avoid T cell recognition; but this increases their recognition by NK cells [69, 70]. However, whether this mechanism applies to NK cell-derived EVs is uncertain, because EVs lack MHC molecules, which prevents the ligation of inhibitory molecules on their surface, leading to NK cell activation. EVs are static, and their function does not depend on the net outcome of activating and inhibitory molecules on their surface.

3.3 Extracellular Vesicles Derived from Cancer Cells

Cancer cells are known for their ability to avoid detection by the immune system, which is commonly referred to as "immune escape" [74]. This is achieved through a range of mechanisms that hinder both the adaptive and innate immune responses. Cancer cells use various inhibitory interactions with immune cells, including direct physical contact and release of soluble factors, to achieve immune evasion [75, 76]. Cancer cell-derived EVs play a significant role in these interactions, exhibiting properties that can either facilitate immune evasion or trigger an immune response (Figure 3.8) [78–80].

In the initial research on metastatic melanoma, it was discovered that the protein tyrosine kinase MET is upregulated in small EVs. These EVs can permanently reprogram circulating bone-marrow progenitor cells into protumorigenic mediators, which facilitate the formation of a premetastatic niche *in vivo* [81]. Another study found that small EVs from pancreatic cancer cells, which contain a large amount of macrophage migration inhibitory factor (MIF), can stimulate Kupffer cells in the liver to release transforming growth factor beta (TGF-β) and remodel the extracellular matrix (ECM) [82]. The recent research has also revealed that cancer cell-derived EVs carry a significant amount of functional mRNA, noncoding RNA, and proteins that are essential for interacting with both the innate and adaptive immune responses [83–85].

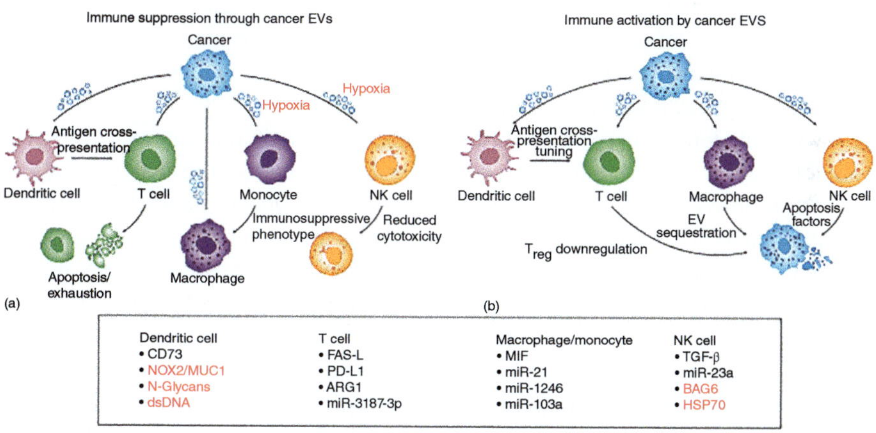

Figure 3.8 Cancer-derived EVs directly regulate tumor progression. (a) Cancer cells release EVs that suppress the function of adaptive and innate immune cells, leading to tumor progression. These EVs inhibit effective antigen cross-presentation in dendritic cells, which contributes to dysfunction in T cells via checkpoint inhibition. Furthermore, cancer EVs drive the polarization of mononuclear cells toward an immunosuppressive phenotype and decrease cytotoxicity in NK cells. (b) Cancer EVs also play a role in tumor recognition. During an immune response, EVs transport foreign antigens to dendritic cells, indirectly activating cytotoxic T cells. Macrophages actively capture cancer-derived EVs from circulation, promoting the recognition of cancer cells. Some molecules carried by EVs stimulate immune activation (shown in red), while others have an immunosuppressive effect (shown in black). Source: Marar et al. [77]/Springer Nature.

Cancer cell-derived EVs play a diverse role in immune suppression, as evidenced by RNA sequencing of small chronic lymphocytic leukemia EVs, which showed that the noncoding Y RNA hY4 increased PD-L1 expression in circulating monocytes via toll-like receptor (TLR) 7 signaling [86]. PD-L1 on small EVs isolated from glioblastoma and metastatic melanomas can directly activate the PD-1–PD-L1 immune checkpoint and prevent T cell activation to prevent autoimmune response [83, 85]. The membrane topology of PD-L1 on small EVs isolated from metastatic melanoma is identical to that of the tumor-cell progenitor, allowing it to functionally bind PD-1. The identification of EVs as carriers of PD-L1 has emerged as a prominent subclass of EV-mediated immune suppression.

In addition, Cancer cell-derived EVs have the ability to transport miRNAs and enzymatically active Arginase-1 (ARG1), which can directly affect T cell activation, proliferation, and cytokine release by downregulating the TCR [84, 87]. For instance, the transfer of miR-498 *in vitro* can directly impact the release of tumor necrosis factor (TNF) in $CD8^+$ T cells in metastatic melanoma. Additionally, miR-3187-3p can suppress CD45 membrane expression and consequent TCR activation in melanoma [87]. ARG1, which is found in ovarian carcinoma EVs, has been identified as a metabolic mechanism for T cell dysfunction. By catalyzing the urea cycle, ARG1-mediated depletion of L-arginine suppresses T cell immune response *in vitro* by downregulating the TCR complex component CD3ζ and causing cell-cycle arrest in the G1 phase via RICTOR in the mTORC2 complex [88, 89]. $ARG1^+$ ovarian cancer EVs, isolated from human plasma and ascites, actively suppress T cell proliferation, both directly and via DC antigen cross-presentation [84]. The potential effects of $ARG1^+$ cancer EVs in T cell inhibition are extensive, as $ARG1^+$ macrophages have been observed to actively arrest T cell proliferation even in the absence of TCR signaling [89].

Growing evidence suggests that the characteristics of the tumor microenvironment play a crucial role in determining the quantity and composition of cancer cell-derived EVs [90–92]. In particular, the hypoxic conditions present in desmoplastic and advanced tumors can increase the release of EVs that are enriched in immunosuppressive proteins and miRNAs [91]. Studies on a mouse model of macrophage infiltration have shown that EVs released by hypoxic B16F0 melanoma cells can promote an anti-inflammatory M2-like phenotype in infiltrating macrophages [93]. These effects are mediated, at least, in part, by miR-103a and Let-7a, which are upregulated under hypoxic conditions and promote M2-like polarization in lung and melanoma cancers by downregulating the insulin-AKT-mTOR signaling pathway [90, 93]. Moreover, hypoxia-induced EVs carrying miR-23a and TGF-β have been implicated in the suppression of NK cells by downregulating CD107a and NKG2D *in vitro* [94]. While chemical and physical cues may also influence the content of EVs, the manner in which they disseminate within tissues could also have significant implications for their immunoregulatory functions.

3.4 Extracellular Vesicles Derived from Plants

In addition to above cell source, recently EVs derived from plants, such as broccoli, ginger, garlic, carrot, grape, blueberry, strawberry, apple, and lemon, has gained great attention due to the potential in various therapies [95].

Plants have been applied for medicinal purposes for a long time, either in their natural form or by extracting active compounds such as polysaccharides, phenols, and terpenoids to inhibit the progression of diseases and repair damages. The recent research focused on plant-derived EVs has shown promising biological activities in various edible plants. These plant-derived EVs have demonstrated anti-inflammatory, anticancer, antibacterial and antioxidation properties. Their efficacy is due to multiple pathways, including gene regulation, intestinal flora, macrophages, gene silencing, and their own active molecules [95]. Although the study of plant-derived EVs is in its early stages, the initial findings are highly encouraging.

3.4.1 Anti-inflammatory

Several studies have demonstrated the remarkable anti-inflammatory properties of plant-derived EVs, such as EVs derived from ginger, grapefruit, and grapes [96–98]. These EVs contain both lipid and RNA components, which may contribute to their anti-inflammatory effects through distinct mechanisms [99]. Notably, the lipid components of plant-derived EVs are crucial for their anti-inflammatory activity. They can regulate gene expression in cells located in inflammatory sites, thereby potentially reducing inflammation. Furthermore, these lipids can modulate the function of macrophages, which are key immune cells involved in inflammation, further contributing to the anti-inflammatory effect of plant-derived EVs [97]. Besides, plant-derived EVs contain various types of RNAs, including miRNAs, which can modify the composition and function of the gut microbiota. These miRNAs are first taken up by the microbiota in the intestine, and then can affect the transcription and translation of target RNAs. This modulation leads to changes in the levels of inflammatory factors or chemokines, resulting in a reduction in inflammation. The use of plant-derived EVs has shown significant anti-inflammatory effects in the treatment of acute and chronic colitis. EVs derived from ginger are known to possess anti-inflammatory properties by activating the translocation of the nuclear factor erythroid 2-related factor 2 (Nrf2) and inducing the production of anti-inflammatory cytokines. On the other hand, EVs derived from grapefruit play a significant role in regulating inflammation by activating the Wnt signaling pathway. This involves the translocation of Nrf2 into the nucleus and the activation of T-cell factor 4 (TCF4) in the Wnt signaling pathway, which collectively contribute to an anti-inflammatory response [100–102].

3.4.2 Anticancer

The research conducted indicates that plant-derived EVs derived from various sources, including citrus lemon, ginger, grape, and grapefruit, exhibit varying levels of therapeutic benefits for a range of cancer types [103, 104]. Furthermore, these EVs are known to have minimal side effects. Specifically, lemon-derived EVs have been found to up-regulate the expression of GADD45a through ROS produced by tumor tissue, leading to S-phase arrest and apoptosis of gastric cancer cells [105]. In addition, EVs derived from citrus, lemon, and grapefruit have also been shown to inhibit the growth of melanoma, lung adenocarcinoma, and breast cancer cell lines to varying degrees [106]. Furthermore, tea flower-derived EVs have demonstrated the ability to accumulate in breast tumors and lung metastatic sites after intravenous injection or oral administration, inhibiting the growth and metastasis of breast cancer and modulating gut microbiota (Figure 3.9) [107].

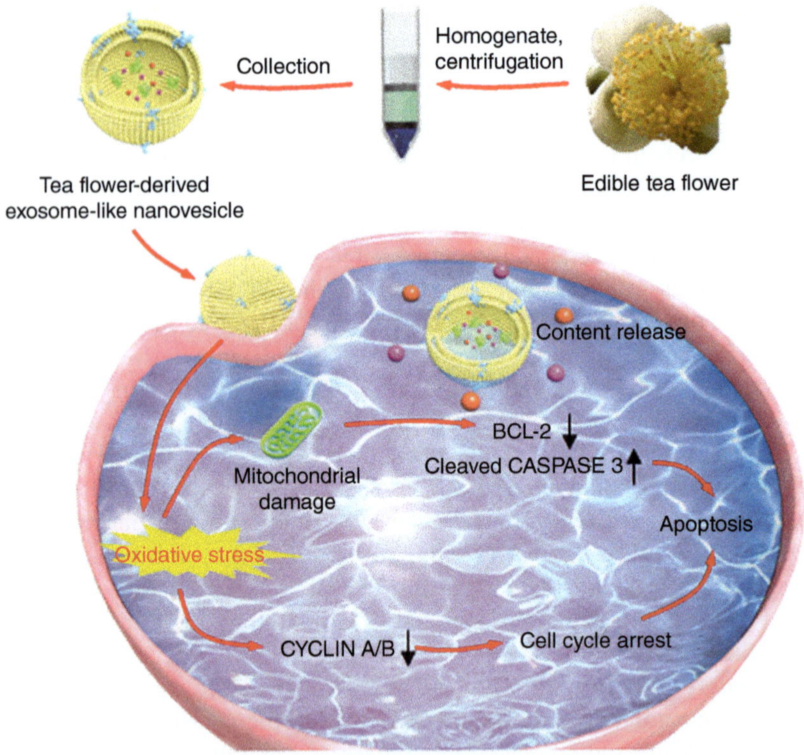

Figure 3.9 Exosome derived from edible tea flowers induced mitochondrial damage, cell cycle arrest, cell apoptosis, microbiota modulation, and further inhibited the growth of breast tumors and their lung metastasis. Source: Chen et al. [107]/with permission from Elsevier.

3.4.3 Antibacterial

Recent research indicates that bacteria can absorb certain plant-derived EVs under specific incubation conditions [108]. This process may serve as a means of regulating bacterial growth via internal microRNAs that stimulate gene expression in the bacteria. For instance, coconut water EVs have been confirmed to facilitate the growth of probiotics like *Lactobacillus plantarum* WCFS1 and MG1655 [109]. Moreover, EVs obtained from Arabidopsis thaliana have the ability to convey small RNAs to the site of fungal infection where they are absorbed by the pathogen *Botrytis cinerea*. These small RNAs can then trigger the silencing of critical fungal genes [110].

3.4.4 Antioxidation

The process of oxidation is responsible for negative outcomes such as aging and inflammation. Plant-derived EVs with a lipid bilayer structure can provide protection to unstable antioxidants in the vesicles. These EVs can transport the antioxidants from fruits and vegetables, which possess significant antioxidant properties, to the plasmids *in vitro*. For instance, lemon-derived EVs are rich in citric acid and vitamin C, which exhibit a remarkable protective effect against oxidative stress in MSCs [111]. Similarly, MSCs can take up EVs derived from strawberries without any adverse effects on their activity. Strawberry-derived EVs can prevent oxidative stress in a dose-dependent manner, potentially due to the high concentration of vitamin C in the vesicles [112]. Therefore, the antioxidant capabilities of plant-derived EVs hold great promise in the fields of medicine and cosmetology.

References

1 Bahmani, L. and Ullah, M. (2022). Different sourced extracellular vesicles and their potential applications in clinical treatments. *Cells* 11 (13).

2 Pittenger, M.F., Mackay, A.M., Beck, S.C. et al. (1999). Multilineage potential of adult human mesenchymal stem cells. *Science* 284 (5411): 143–147.

3 Phinney, D.G. and Prockop, D.J. (2007). Concise review: mesenchymal stem/multipotent stromal cells: the state of transdifferentiation and modes of tissue repair–current views. *Stem Cells* 25 (11): 2896–2902.

4 Timmermans, F., Plum, J., Yoder, M.C. et al. (2009). Endothelial progenitor cells: identity defined? *J. Cell. Mol. Med.* 13 (1): 87–102.

5 Yoder, M.C. (2012). Human endothelial progenitor cells. *Cold Spring Harbor Perspect. Med.* 2 (7): a006692.

6 Henning, R.J. (2011). Stem cells in cardiac repair. *Future Cardiol.* 7 (1): 99–117.

7 Parmacek, M.S. (2006). Cardiac stem cells and progenitors: developmental biology and therapeutic challenges. *Trans. Am. Clin. Climatol. Assoc.* 117: 239–255; discussion 55–56.

8 Clarke, D.L., Johansson, C.B., Wilbertz, J. et al. (2000). Generalized potential of adult neural stem cells. *Science* 288 (5471): 1660–1663.

9 Zhang, B., Yeo, R.W., Tan, K.H., and Lim, S.K. (2016). Focus on extracellular vesicles: therapeutic potential of stem cell-derived extracellular vesicles. *Int. J. Mol. Sci.* 17 (2): 174.

10 Andaloussi, S.E.L., Mager, I., Breakefield, X.O., and Wood, M.J. (2013). Extracellular vesicles: biology and emerging therapeutic opportunities. *Nat. Rev. Drug Discovery* 12 (5): 347–357.

11 Ghannam, S., Bouffi, C., Djouad, F. et al. (2010). Immunosuppression by mesenchymal stem cells: mechanisms and clinical applications. *Stem Cell Res. Ther.* 1 (1): 2.

12 Zhao, A.G., Shah, K., Cromer, B., and Sumer, H. (2020). Mesenchymal stem cell-derived extracellular vesicles and their therapeutic potential. *Stem Cells Int.* 2020: 8825771.

13 Katsuda, T., Kosaka, N., Takeshita, F., and Ochiya, T. (2013). The therapeutic potential of mesenchymal stem cell-derived extracellular vesicles. *Proteomics* 13 (10–11): 1637–1653.

14 Humphreys, B.D. and Bonventre, J.V. (2008). Mesenchymal stem cells in acute kidney injury. *Annu. Rev. Med.* 59: 311–325.

15 Bruno, S., Grange, C., Deregibus, M.C. et al. (2009). Mesenchymal stem cell-derived microvesicles protect against acute tubular injury. *J. Am. Soc. Nephrol.* 20 (5): 1053–1067.

16 Bruno, S., Grange, C., Collino, F. et al. (2012). Microvesicles derived from mesenchymal stem cells enhance survival in a lethal model of acute kidney injury. *PLoS One* 7 (3): e33115.

17 Gatti, S., Bruno, S., Deregibus, M.C. et al. (2011). Microvesicles derived from human adult mesenchymal stem cells protect against ischaemia-reperfusion-induced acute and chronic kidney injury. *Nephrol. Dial. Transplant.* 26 (5): 1474–1483.

18 Timmers, L., Lim, S.K., Arslan, F. et al. (2007). Reduction of myocardial infarct size by human mesenchymal stem cell conditioned medium. *Stem Cell Res. Ther.* 1 (2): 129–137.

19 Lai, R.C., Arslan, F., Lee, M.M. et al. (2010). Exosome secreted by MSC reduces myocardial ischemia/reperfusion injury. *Stem Cell Res. Ther.* 4 (3): 214–222.

20 Lai, R.C., Arslan, F., Tan, S.S. et al. (2010). Derivation and characterization of human fetal MSCs: an alternative cell source for large-scale production of cardioprotective microparticles. *J. Mol. Cell. Cardiol.* 48 (6): 1215–1224.

21 Zhang, N., Song, Y., Huang, Z. et al. (2020). Monocyte mimics improve mesenchymal stem cell-derived extracellular vesicle homing in a mouse MI/RI model. *Biomaterials* 255: 120168.

22 Lee, J.K., Park, S.R., Jung, B.K. et al. (2013). Exosomes derived from mesenchymal stem cells suppress angiogenesis by down-regulating VEGF expression in breast cancer cells. *PLoS One* 8 (12): e84256.

23 Katakowski, M., Buller, B., Zheng, X. et al. (2013). Exosomes from marrow stromal cells expressing miR-146b inhibit glioma growth. *Cancer Lett.* 335 (1): 201–204.

24 Ma, B., Jiang, H., Jia, J. et al. (2012). Murine bone marrow stromal cells pulsed with homologous tumor-derived exosomes inhibit proliferation of liver cancer cells. *Clin. Transl. Oncol.* 14 (10): 764–773.

25 McKay, R. (1997). Stem cells in the central nervous system. *Science* 276 (5309): 66–71.

26 Gage, F.H. (2000). Mammalian neural stem cells. *Science* 287 (5457): 1433–1438.

27 Hering, C. and Shetty, A.K. (2023). Extracellular vesicles derived from neural stem cells, astrocytes, and microglia as therapeutics for easing TBI-induced brain dysfunction. *Stem Cells Transl. Med.* 12 (3): 140–153.

28 Colombo, M., Raposo, G., and Thery, C. (2014). Biogenesis, secretion, and intercellular interactions of exosomes and other extracellular vesicles. *Annu. Rev. Cell Dev. Biol.* 30: 255–289.

29 Zhang, L., Liu, H., Jia, L. et al. (2019). Exosomes mediate hippocampal and cortical neuronal injury induced by hepatic ischemia-reperfusion injury through activating pyroptosis in rats. *Oxid. Med. Cell. Longevity* 2019: 3753485.

30 Han, Y., Chu, X., Cui, L. et al. (2020). Neuronal mitochondria-targeted therapy for Alzheimer's disease by systemic delivery of resveratrol using dual-modified novel biomimetic nanosystems. *Drug Deliv.* 27 (1): 502–518.

31 Chen, W., Zheng, P., Hong, T. et al. (2020). Astrocytes-derived exosomes induce neuronal recovery after traumatic brain injury via delivering gap junction alpha 1-20 K. *J. Tissue Eng. Regener. Med.* 14 (3): 412–423.

32 Shi, M., Sheng, L., Stewart, T. et al. (2019). New windows into the brain: central nervous system-derived extracellular vesicles in blood. *Prog. Neurobiol.* 175: 96–106.

33 Soares Martins, T., Trindade, D., Vaz, M. et al. (2021). Diagnostic and therapeutic potential of exosomes in Alzheimer's disease. *J. Neurochem.* 156 (2): 162–181.

34 Sun, M.K., Passaro, A.P., Latchoumane, C.F. et al. (2020). Extracellular vesicles mediate neuroprotection and functional recovery after traumatic brain injury. *J. Neurotrauma* 37 (11): 1358–1369.

35 Laine, S.K., Hentunen, T., and Laitala-Leinonen, T. (2012). Do microRNAs regulate bone marrow stem cell niche physiology? *Gene* 497 (1): 1–9.

36 Deregibus, M.C., Cantaluppi, V., Calogero, R. et al. (2007). Endothelial progenitor cell derived microvesicles activate an angiogenic program in endothelial cells by a horizontal transfer of mRNA. *Blood* 110 (7): 2440–2448.

37 Cantaluppi, V., Gatti, S., Medica, D. et al. (2012). Microvesicles derived from endothelial progenitor cells protect the kidney from ischemia-reperfusion injury by microRNA-dependent reprogramming of resident renal cells. *Kidney Int.* 82 (4): 412–427.

38 Ranghino, A., Cantaluppi, V., Grange, C. et al. (2012). Endothelial progenitor cell-derived microvesicles improve neovascularization in a murine model of hindlimb ischemia. *Int. J. Immunopathol. Pharmacol.* 25 (1): 75–85.

39 Wang, J., Chen, S., Ma, X. et al. (2013). Effects of endothelial progenitor cell-derived microvesicles on hypoxia/reoxygenation-induced endothelial dysfunction and apoptosis. *Oxid. Med. Cell. Longevity* 2013: 572729.

40 Gu, S., Zhang, W., Chen, J. et al. (2014). EPC-derived microvesicles protect cardiomyocytes from Ang II-induced hypertrophy and apoptosis. *PLoS One* 9 (1): e85396.

41 Matsuura, K., Honda, A., Nagai, T. et al. (2009). Transplantation of cardiac progenitor cells ameliorates cardiac dysfunction after myocardial infarction in mice. *J. Clin. Invest.* 119 (8): 2204–2217.

42 Tang, X.L., Rokosh, G., Sanganalmath, S.K. et al. (2010). Intracoronary administration of cardiac progenitor cells alleviates left ventricular dysfunction in rats with a 30-day-old infarction. *Circulation* 121 (2): 293–305.

43 Chen, L., Wang, Y., Pan, Y. et al. (2013). Cardiac progenitor-derived exosomes protect ischemic myocardium from acute ischemia/reperfusion injury. *Biochem. Biophys. Res. Commun.* 431 (3): 566–571.

44 Barile, L., Lionetti, V., Cervio, E. et al. (2014). Extracellular vesicles from human cardiac progenitor cells inhibit cardiomyocyte apoptosis and improve cardiac function after myocardial infarction. *Cardiovasc. Res.* 103 (4): 530–541.

45 Gray, W.D., French, K.M., Ghosh-Choudhary, S. et al. (2015). Identification of therapeutic covariant microRNA clusters in hypoxia-treated cardiac progenitor cell exosomes using systems biology. *Circ. Res.* 116 (2): 255–263.

46 Ong, S.G., Lee, W.H., Huang, M. et al. (2014). Cross talk of combined gene and cell therapy in ischemic heart disease: role of exosomal microRNA transfer. *Circulation* 130 (11 Suppl 1): S60–S69.

47 Ratajczak, J., Miekus, K., Kucia, M. et al. (2006). Embryonic stem cell-derived microvesicles reprogram hematopoietic progenitors: evidence for horizontal transfer of mRNA and protein delivery. *Leukemia* 20 (5): 847–856.

48 Herrera, M.B., Fonsato, V., Gatti, S. et al. (2010). Human liver stem cell-derived microvesicles accelerate hepatic regeneration in hepatectomized rats. *J. Cell. Mol. Med.* 14 (6B): 1605–1618.

49 Sahoo, S., Klychko, E., Thorne, T. et al. (2011). Exosomes from human CD34(+) stem cells mediate their proangiogenic paracrine activity. *Circ. Res.* 109 (7): 724–728.

50 Veerman, R.E., Gucluler Akpinar, G., Eldh, M., and Gabrielsson, S. (2019). Immune cell-derived extracellular vesicles - functions and therapeutic applications. *Trends Mol. Med.* 25 (5): 382–394.

51 Wen, C., Seeger, R.C., Fabbri, M. et al. (2017). Biological roles and potential applications of immune cell-derived extracellular vesicles. *J. Extracell. Vesicles* 6 (1): 1400370.

52 Shapouri-Moghaddam, A., Mohammadian, S., Vazini, H. et al. (2018). Macrophage plasticity, polarization, and function in health and disease. *J. Cell. Physiol.* 233 (9): 6425–6440.

53 Wu, Q., Fu, S., Xiao, H. et al. (2023). Advances in extracellular vesicle nanotechnology for precision theranostics. *Adv. Sci. (Weinh)* 10 (3): e2204814.

54 Gunassekaran, G.R., Poongkavithai Vadevoo, S.M., Baek, M.C., and Lee, B. (2021). M1 macrophage exosomes engineered to foster M1 polarization and target the IL-4 receptor inhibit tumor growth by reprogramming tumor-associated macrophages into M1-like macrophages. *Biomaterials* 278: 121137.

55 Li, H., Feng, Y., Zheng, X. et al. (2022). M2-type exosomes nanoparticles for rheumatoid arthritis therapy via macrophage re-polarization. *J. Controlled Release* 341: 16–30.

56 Essandoh, K., Li, Y., Huo, J., and Fan, G.C. (2016). MiRNA-mediated macrophage polarization and its potential role in the regulation of inflammatory response. *Shock* 46 (2): 122–131.

57 Tang, T.T., Wang, B., Wu, M. et al. (2020). Extracellular vesicle-encapsulated IL-10 as novel nanotherapeutics against ischemic AKI. *Sci. Adv.* 6 (33): eaaz0748.

58 Escudier, B., Dorval, T., Chaput, N. et al. (2005). Vaccination of metastatic melanoma patients with autologous dendritic cell (DC) derived-exosomes: results of thefirst phase I clinical trial. *J. Transl. Med.* 3 (1): 10.

59 Morse, M.A., Garst, J., Osada, T. et al. (2005). A phase I study of dexosome immunotherapy in patients with advanced non-small cell lung cancer. *J. Transl. Med.* 3 (1): 9.

60 Gehrmann, U., Hiltbrunner, S., Georgoudaki, A.M. et al. (2013). Synergistic induction of adaptive antitumor immunity by codelivery of antigen with alpha-galactosylceramide on exosomes. *Cancer Res.* 73 (13): 3865–3876.

61 Lu, Z., Zuo, B., Jing, R. et al. (2017). Dendritic cell-derived exosomes elicit tumor regression in autochthonous hepatocellular carcinoma mouse models. *J. Hepatol.* 67 (4): 739–748.

62 Ellwanger, J.H., Crovella, S., Dos Reis, E.C. et al. (2016). Exosomes are possibly used as a tool of immune regulation during the dendritic cell-based immune therapy against HIV-I. *Med. Hypotheses.* 95: 67–70.

63 Zhang, Y., Cai, Z., Shen, Y. et al. (2021). Hydrogel-load exosomes derived from dendritic cells improve cardiac function via Treg cells and the polarization of macrophages following myocardial infarction. *J. Nanobiotechnol.* 19 (1): 271.

64 Blanchard, N., Lankar, D., Faure, F. et al. (2002). TCR activation of human T cells induces the production of exosomes bearing the TCR/CD3/zeta complex. *J. Immunol.* 168 (7): 3235–3241.

65 Torralba, D., Baixauli, F., Villarroya-Beltri, C. et al. (2018). Priming of dendritic cells by DNA-containing extracellular vesicles from activated T cells through antigen-driven contacts. *Nat. Commun.* 9 (1): 2658.

66 Seo, N., Shirakura, Y., Tahara, Y. et al. (2018). Activated CD8(+) T cell extracellular vesicles prevent tumour progression by targeting of lesional mesenchymal cells. *Nat. Commun.* 9 (1): 435.

67 Yu, X., Huang, C., Song, B. et al. (2013). CD4+CD25+ regulatory T cells-derived exosomes prolonged kidney allograft survival in a rat model. *Cell Immunol.* 285 (1–2): 62–68.

68 Long, E.O. (2002). Tumor cell recognition by natural killer cells. *Semin. Cancer Biol.* 12 (1): 57–61.

69 Lugini, L., Cecchetti, S., Huber, V. et al. (2012). Immune surveillance properties of human NK cell-derived exosomes. *J. Immunol.* 189 (6): 2833–2842.

70 Jong, A.Y., Wu, C.H., Li, J. et al. (2017). Large-scale isolation and cytotoxicity of extracellular vesicles derived from activated human natural killer cells. *J. Extracell. Vesicles* 6 (1): 1294368.

71 Zhu, L., Oh, J.M., Gangadaran, P. et al. (2018). Targeting and therapy of glioblastoma in a mouse model using exosomes derived from natural killer cells. *Front. Immunol.* 9: 824.

72 Zhu, L., Kalimuthu, S., Gangadaran, P. et al. (2017). Exosomes derived from natural killer cells exert therapeutic effect in melanoma. *Theranostics* 7 (10): 2732–2745.

73 Watson, D.C., Bayik, D., Srivatsan, A. et al. (2016). Efficient production and enhanced tumor delivery of engineered extracellular vesicles. *Biomaterials* 105: 195–205.

74 Becker, A., Thakur, B.K., Weiss, J.M. et al. (2016). Extracellular vesicles in cancer: cell-to-cell mediators of metastasis. *Cancer Cell.* 30 (6): 836–848.

75 Hiratsuka, S., Watanabe, A., Aburatani, H., and Maru, Y. (2006). Tumour-mediated upregulation of chemoattractants and recruitment of myeloid cells predetermines lung metastasis. *Nat. Cell Biol.* 8 (12): 1369–1375.

76 Oskarsson, T., Acharyya, S., Zhang, X.H. et al. (2011). Breast cancer cells produce tenascin C as a metastatic niche component to colonize the lungs. *Nat. Med.* 17 (7): 867–874.

77 Marar, C., Starich, B., and Wirtz, D. (2021). Extracellular vesicles in immunomodulation and tumor progression. *Nat. Immunol.* 22 (5): 560–570.

78 Hoshino, A., Costa-Silva, B., Shen, T.L. et al. (2015). Tumour exosome integrins determine organotropic metastasis. *Nature* 527 (7578): 329–335.

79 Clayton, A. and Tabi, Z. (2005). Exosomes and the MICA-NKG2D system in cancer. *Blood Cells Mol. Dis.* 34 (3): 206–213.

80 Capello, M., Vykoukal, J.V., Katayama, H. et al. (2019). Exosomes harbor B cell targets in pancreatic adenocarcinoma and exert decoy function against complement-mediated cytotoxicity. *Nat. Commun.* 10 (1): 254.

81 Peinado, H., Aleckovic, M., Lavotshkin, S. et al. (2012). Melanoma exosomes educate bone marrow progenitor cells toward a pro-metastatic phenotype through MET. *Nat. Med.* 18 (6): 883–891.

82 Costa-Silva, B., Aiello, N.M., Ocean, A.J. et al. (2015). Pancreatic cancer exosomes initiate pre-metastatic niche formation in the liver. *Nat. Cell Biol.* 17 (6): 816–826.

83 Chen, G., Huang, A.C., Zhang, W. et al. (2018). Exosomal PD-L1 contributes to immunosuppression and is associated with anti-PD-1 response. *Nature* 560 (7718): 382–386.

84 Czystowska-Kuzmicz, M., Sosnowska, A., Nowis, D. et al. (2019). Small extracellular vesicles containing arginase-1 suppress T-cell responses and promote tumor growth in ovarian carcinoma. *Nat. Commun.* 10 (1): 3000.

85 Ricklefs, F.L., Alayo, Q., Krenzlin, H. et al. (2018). Immune evasion mediated by PD-L1 on glioblastoma-derived extracellular vesicles. *Sci. Adv.* 4 (3): eaar2766.

86 Haderk, F., Schulz, R., Iskar, M. et al. (2017). Tumor-derived exosomes modulate PD-L1 expression in monocytes. *Sci. Immunol.* 2 (13).

87 Vignard, V., Labbe, M., Marec, N. et al. (2020). MicroRNAs in tumor exosomes drive immune escape in melanoma. *Cancer Immunol. Res.* 8 (2): 255–267.

88 Rodriguez, P.C., Zea, A.H., Culotta, K.S. et al. (2002). Regulation of T cell receptor CD3zeta chain expression by L-arginine. *J. Biolumin. Chemilumin.* 277 (24): 21123–21129.

89 Van de Velde, L.A., Subramanian, C., Smith, A.M. et al. (2017). T cells encountering myeloid cells programmed for amino acid-dependent immunosuppression use rictor/mTORC2 protein for proliferative checkpoint decisions. *J. Biolumin. Chemilumin.* 292 (1): 15–30.

90 Hsu, Y.L., Hung, J.Y., Chang, W.A. et al. (2018). Hypoxic lung-cancer-derived extracellular vesicle MicroRNA-103a increases the oncogenic effects of macrophages by targeting PTEN. *Mol. Ther.* 26 (2): 568–581.

91 de Jong, O.G., Verhaar, M.C., Chen, Y. et al. (2012). Cellular stress conditions are reflected in the protein and RNA content of endothelial cell-derived exosomes. *J. Extracell. Vesicles* 1.

92 Lenzini, S., Bargi, R., Chung, G., and Shin, J.W. (2020). Matrix mechanics and water permeation regulate extracellular vesicle transport. *Nat. Nanotechnol.* 15 (3): 217–223.

93 Park, J.E., Dutta, B., Tse, S.W. et al. (2019). Hypoxia-induced tumor exosomes promote M2-like macrophage polarization of infiltrating myeloid cells and microRNA-mediated metabolic shift. *Oncogene* 38 (26): 5158–5173.

94 Berchem, G., Noman, M.Z., Bosseler, M. et al. (2016). Hypoxic tumor-derived microvesicles negatively regulate NK cell function by a mechanism involving TGF-beta and miR23a transfer. *Oncoimmunology* 5 (4): e1062968.

95 Wang, Y., Wang, J., Ma, J. et al. (2022). Focusing on future applications and current challenges of plant derived extracellular vesicles. *Pharmaceuticals (Basel)* 15 (6).

96 Ju, S., Mu, J., Dokland, T. et al. (2013). Grape exosome-like nanoparticles induce intestinal stem cells and protect mice from DSS-induced colitis. *Mol. Ther.* 21 (7): 1345–1357.

97 Wang, B., Zhuang, X., Deng, Z.B. et al. (2014). Targeted drug delivery to intestinal macrophages by bioactive nanovesicles released from grapefruit. *Mol. Ther.* 22 (3): 522–534.

98 Kim, J., Li, S., Zhang, S., and Wang, J. (2022). Plant-derived exosome-like nanoparticles and their therapeutic activities. *Asian J. Pharm. Sci.* 17 (1): 53–69.

99 Suharta, S., Barlian, A., Hidajah, A.C. et al. (2021). Plant-derived exosome-like nanoparticles: a concise review on its extraction methods, content, bioactivities, and potential as functional food ingredient. *J. Food Sci.* 86 (7): 2838–2850.

100 Kim, S.W., Lee, H.K., Shin, J.H., and Lee, J.K. (2013). Up-down regulation of HO-1 and iNOS gene expressions by ethyl pyruvate via recruiting p300 to Nrf2 and depriving It from p65. *Free Radical Biol. Med.* 65: 468–476.

101 Luo, C., Urgard, E., Vooder, T., and Metspalu, A. (2011). The role of COX-2 and Nrf2/ARE in anti-inflammation and antioxidative stress: aging and anti-aging. *Med. Hypotheses.* 77 (2): 174–178.

102 Gersemann, M., Stange, E.F., and Wehkamp, J. (2011). From intestinal stem cells to inflammatory bowel diseases. *World J. Gastroenterol.* 17 (27): 3198–3203.

103 Cirmi, S., Maugeri, A., Ferlazzo, N. et al. (2017). Anticancer potential of citrus juices and their extracts: a systematic review of both preclinical and clinical studies. *Front. Pharmacol.* 8: 420.

104 Zhang, L., He, F., Gao, L. et al. (2021). Engineering exosome-like nanovesicles derived from asparagus cochinchinensis can inhibit the proliferation of hepatocellular carcinoma cells with better safety profile. *Int. J. Nanomed.* 16: 1575–1586.

105 Yang, M., Liu, X., Luo, Q. et al. (2020). An efficient method to isolate lemon derived extracellular vesicles for gastric cancer therapy. *J. Nanobiotechnol.* 18 (1): 100.

106 Stanly, C., Alfieri, M., Ambrosone, A. et al. (2020). Grapefruit-derived micro and nanovesicles show distinct metabolome profiles and anticancer activities in the A375 human melanoma cell line. *Cells* 9 (12).

107 Chen, Q., Li, Q., Liang, Y. et al. (2022). Natural exosome-like nanovesicles from edible tea flowers suppress metastatic breast cancer via ROS generation and microbiota modulation. *Acta Pharmacol. Sin. B* 12 (2): 907–923.

108 Sundaram, K., Miller, D.P., Kumar, A. et al. (2019). Plant-derived exosomal nanoparticles inhibit pathogenicity of porphyromonas gingivalis. *iScience* 21: 308–327.

109 Yu, S., Zhao, Z., Xu, X. et al. (2019). Characterization of three different types of extracellular vesicles and their impact on bacterial growth. *Food Chem.* 272: 372–378.

110 Cai, Q., Qiao, L., Wang, M. et al. (2018). Plants send small RNAs in extracellular vesicles to fungal pathogen to silence virulence genes. *Science* 360 (6393): 1126–1129.

111 Baldini, N., Torreggiani, E., Roncuzzi, L. et al. (2018). Exosome-like nanovesicles isolated from citrus limon L. exert antioxidative effect. *Curr. Pharm. Biotechnol.* 19 (11): 877–885.

112 Perut, F., Roncuzzi, L., Avnet, S. et al. (2021). Strawberry-derived exosome-like nanoparticles prevent oxidative stress in human mesenchymal stromal cells. *Biomolecules* 11 (1).

4

Biomedical Applications of Extracellular Vesicles in Treatment of Disease

Fei Wang[1,2], *Jiacong Ai*[1,2], *Ziyang Zhang*[3], *Yuanhang Li*[3], *and Zhenhua Li*[1,2]

[1] *The Tenth Affiliated Hospital of Southern Medical University, 78 Wanjiang Avenue, Dongguan, Guangdong 523059, China*
[2] *Guangdong Provincial Key Laboratory of Shock and Microcirculation, Shatai South Avenue, Guangzhou, Guangdong 510080, China*
[3] *Hebei University, College of Pharmaceutical Science, Key Laboratory of Pharmaceutical Quality Control of Hebei Province, Wusi East Avenue, Baoding, Hebei 071002, China*

Extracellular vesicles (EVs) represent a class of natural nanoparticles that are secreted by almost all cells in both normal and pathological states. Thus, EVs are widely found in tissues, serum, or other body fluids [1]. In the past decades, abundant studies on EVs have broadened our understanding of their biological, chemical, and physical characteristics and functions, promoting the biomedical applications of EV-based therapeutics for disease treatment [2–4]. For example, the natural capability of EVs to transport substances renders them ideal drug carriers. Accordingly, EVs have been widely investigated as delivery systems of various therapeutics, such as nucleic acids, peptides, and small-molecule drugs [5, 6]. Nevertheless, the clinical translation of EV-based therapy suffers from a few issues, such as insufficient targeting and specificity, limited retention, and poor drug loading efficiency [7, 8]. Thus, various techniques, including genetic engineering, chemical modification, and membrane fusion technology, have been proposed to enhance the therapeutic effects of EVs [9, 10]. Meanwhile, different types of biomaterial-based platforms, such as hydrogels, scaffolds, and microneedles, are developed to further improve the efficacy and reduce the side effects of EVs [11, 12]. In this chapter, we will focus on the biomedical applications of EVs over the past ten years. In particular, the recent achievements of EVs and artificial nanovesicles in the treatment of major diseases, including cardiovascular diseases, tissue engineering and regenerative medicine, cancers, respiratory diseases, and metabolic diseases, will be summarized. This chapter will also discuss the challenges and corresponding solutions in the development of EV-based therapeutic agents. Other reviews or chapters are suggested as references for readers to get a complete picture of EV-based disease therapy [13–35].

Biomedical Applications of Extracellular Vesicles, First Edition.
Edited by Zhenhua Li, Xing-Jie Liang, and Ke Cheng.

4.1 Tissue Engineering and Regenerative Medicine

Mesenchymal stem cells (MSC) repair damaged tissues mainly through paracrine [36]. One of the important secretory paracrine components of MSCs is extracellular vesicles (EVs) [37]. Thus, EVs have been considered and widely applied as a promising strategy for tissue regeneration, and the efficacy of EVs highly depends on their source, isolation method, administration method, dosage, and targeted tissue [38].

For example, various EV-based techniques have been developed for bone regeneration. Osteoporosis is a systematic bone disease that occurs when bone destruction exceeds new bone formation [39]. Osteoporosis is characterized by low bone mass and improvement of bone microstructure [40], leading to increased bone brittleness and risk of fracture. Thus, the prevention treatment of osteoporosis usually falls to therapeutic strategies to inhibit osteoclast bone resorption and promote bone formation. Local administration of exosomes derived from human-induced pluripotent stem cells (HIPSCs) and bone marrow MSCs could effectively repair bone defects in ovariectomized rats via the promotion of osteogenesis and angiogenesis [41]. Nevertheless, isolating MSCs from bone marrow is a painful and invasive process [42]. It is beneficial to find a simple, safe, and convenient source of stem cells to collect EVs for bone regeneration. To this end, Chen et al. proposed human urine-derived stem cells (USCs) as the source to isolate EVs (USC-EVs), as urine is unlimited and easily available, and could be noninvasively collected [43]. The authors further showed that the systemic administration of USC-EVs could effectively reduce bone loss and maintain bone strength in osteoporotic animals via the promotion of osteoblastic bone formation and inhibition of osteoclastic bone resorption. More importantly, the results indicated that the antiosteoporotic features of USC-EVs are not influenced by factors, such as age, sex, and health status of USC donors. Mechanistic investigation showed that the USC-EVs were rich in collagen triple helix repeat 1 (CTHRC1) and osteoprotegerin (OPG) proteins, which is essential for pro-osteogenic and antiosteoclastic effects. This work highlighted the therapeutic potential of human urine-derived stem cells (USCs)-derived EVs in the treatment of osteoporosis via osteogenesis promotion and osteoclastogenesis suppression [43].

Very recently, Huang et al. used skeletal muscle tissue-derived EVs (Mu-EVs) as a cell-free therapy for the treatment of discarded osteoporosis (Figure 4.1) [44]. The results indicated that Mu-EVs, which possessed the typical characteristics of EVs, could be isolated from skeletal muscle tissue easily and plentifully. *In vitro*, experiments showed that Mu-EVs harvested from normal skeletal muscle were subjected to phagocytization of bone marrow stem cells (BMSCs) and osteoclasts (OCs), subsequently promoting osteogenic differentiation of BMSCs and inhibiting formation of OCs. Therefore, Mu-EVs weakened the osteogenic effects of BMSCs and enhanced the osteoclast effects of monocytes. *In vivo*, results indicated that Mu-EVs could validly reverse osteoporosis via the promotion of bone formation and inhibition of bone resorption. Their work suggested that the Mu-EVs based therapy is a potential cell-free therapy for the treatment of osteoporosis [44].

In another study, Ma et al. came up with a novel concept called "morphological memory of small extracellular vesicles (sEVs)" for bone regeneration

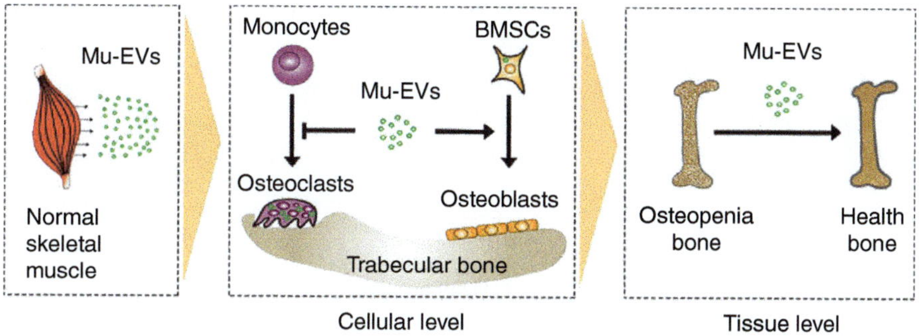

Figure 4.1 Schematic demonstration of osteoporosis treatment mediated by Mu-EVs. Source: Huang et al. [44]/with permission of Elsevier.

(Figure 4.2) [45]. The authors conducted nanomorphology on titanium plates that were treated with alkali and heat (Ti8) to facilitate the differentiation of human mesenchymal stem cells (hMSC). After incubation of hMSC on a nanomorphology Ti plate for 21 days, the author isolated sEVs from these hMSCs (Ti8-21-sEV). The *in vitro* and *in vivo* experiments demonstrated that the Ti8-21-sEV possessed superior pro-osteogenic ability. Single-cell RNA sequencing further indicated that the promotion of Ti8-21-sEV on bone regeneration was associated with osteogenesis-related pathways, including PI3K-AKT signal pathway, MAPK signal pathway, focal adhesion, and extracellular matrix-receptor interaction. Finally, the authors modified Ti8-21-sEV on a 3D printed porous polyetheretherketone (PEEK) scaffold. A rabbit femoral condylar defect model was utilized to confirm the optimal bone ingrowth effect of the Ti8-21-sEVs. This work proved the memory function of Ti8-21-sEVs via a copy of bone-promoting information of nanomorphology, encouraging the development of other types of sEVs with morphology memory for tissue regeneration applications [45].

Recently, EVs derived from stem cells have been increasingly used as bioactive materials for tissue regeneration to accelerate wound healing or reduce scar formation [46]. Although these cells are natural and can suppress inflammation, biosafety issues (such as immunogenicity) and limited production should be taken into account when using stem cells or stem cell-derived EVs in clinical applications [8]. As alternative biologically stable materials that can be used to develop more effective therapeutics, milk-derived EV has attracted increased attention. Milk-derived EVs indicate the advantages of excellent biosafety, high yield, and negligible affection on the host immune system [47]. In addition, milk is a relatively economical and scalable source for EV separation [48]. For these reasons, Yang et al. harvested bovine milk EV released from cow mammary epithelial cells and explored their potential to promote wound healing (Figure 4.3) [49]. The data indicated that milk EVs promoted the proliferation and migration of fibroblast cells, as well as formation of endothelial tube. Particularly, colostrum-derived EVs (Colos EVs) facilitated cell proliferation, tissue remodeling, and angiogenesis. Moreover, Colos EVs significantly promoted re-epithelialization, activation of angiogenesis and maturation of extracellular matrix (ECM) in an excisional wound mouse model. The authors

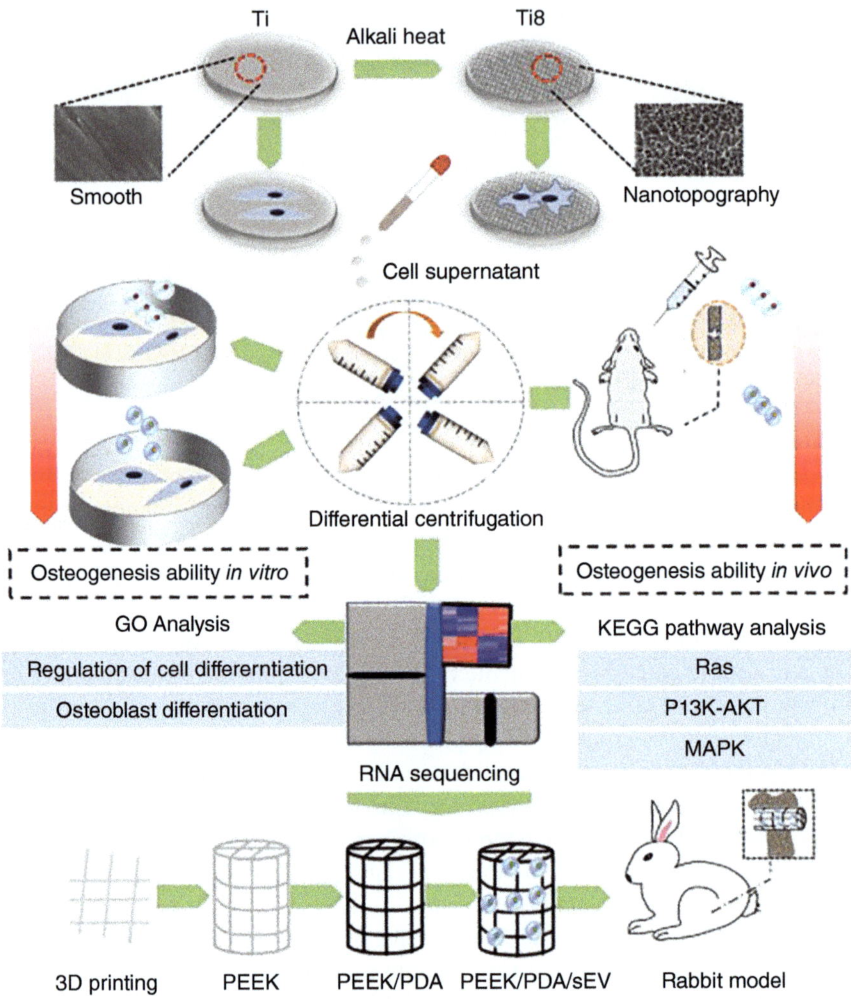

Figure 4.2 Development of small EVs with nanomorphology memory for the promotion of osteogenesis. Source: Ma et al. [45]/with permission from Elsevier.

further investigated the underlying mechanisms of the wound-healing ability of milk EVs. The results showed that "cocktails" of both anti-inflammatory cytokines and ECM remodeling factors were provided in milk EVs. The data also suggested that wound healing induced by milk EVs may be regulated by TGF-β/Smad signal transduction. The proteomic data provided convincing evidence that milk EV-based therapy may be an attractive approach to promote wound healing. This work on the superior activity of milk EVs indicated that milk EVs hold great promise as an anti-inflammatory therapeutic, especially for cutaneous wounds [49].

Renal tissue engineering and regeneration techniques provide great potential for the treatment of chronic kidney diseases [50]. However, the complexity of renal tissue brings critical challenges to the application and clinical translation of

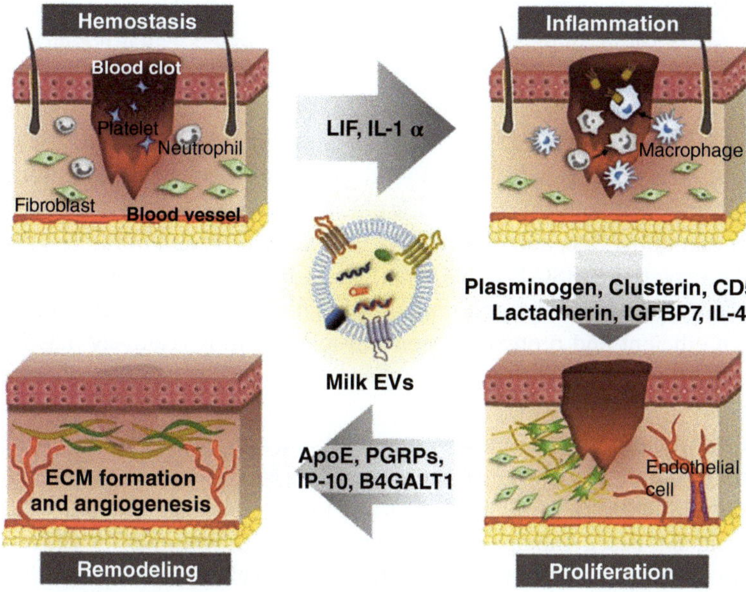

Figure 4.3 Schematic illumination of milk EVs-mediated anti-inflammatory wound healing. Source: Kim et al. [49]/John Wiley & Sons.

regenerative medicine in renal tissue regeneration. To this end, Ko et al. proposed a novel porous pneumatic microextrusion (PME)—composite scaffold for renal tissue regeneration and function maintenance [51]. In this work, poly(lactic acid–glycolic acid copolymer) (PLGA), magnesium hydroxide (MH), and acellular porcine kidney extracellular matrix (kECM) were used to construct the scaffolds. Afterward, polydeoxyribonucleotide (PDRN) and TNF-α/IFN-γ-primed MSC-derived EVs (TI-EVs) were used to functionalize the PME scaffold to enhance the regeneration and maintenance of a functional renal organ. The results indicated remarkable synergistic effects of PDRN and TI-EVs in tissue regeneration with the aspects of cell proliferation, angiogenesis, fibrosis, and inflammation. Also, the interaction of MH, kECM, PDRN, and TI-EV was demonstrated to offer an optimal microenvironment for renal cell morphogenesis. In addition, the developed scaffold induced superior therapeutic efficacy than the available PME scaffold stents in the regeneration of glomerulus and recovery of renal function in a nephrectomy animal model. Taken together, the proposed PME scaffold represents a novel advanced platform for tissue engineering and regeneration [51].

Although the therapeutic applications of MSC-EVs have been demonstrated in many preclinical investigations of heart diseases, the clinical translation of EV-based drugs is still restricted, mainly by the sufficient production of EVs [52]. To this end, Wang et al. proposed a novel approach for the massive generation of nanovesicles (NVs) with therapeutic efficacy from MSC cells via scale-up extrusion [53]. The extruded MSC-NVs could be considered the tiny version of native MSC-EVs about the expression of surface markers. What is existing is that the yield of MSC-NVs is relatively >20 times that of native MSC-EVs. After careful investigations, the authors indicated that the MSC-NVs possess equal potency to native MSC-EVs in cardiac repair. Briefly, the *in vitro* studies demonstrated the protective effects and therapeutic potential of MSC-NVs on cardiac functions. After that, the authors injected MSC-NVs intramyocardially in the injured heart in an ischemia–reperfusion model, and found decreased scar size and cardiac function preservation. Of note, the therapeutic efficacy of MSC-NVs was found to be comparable to that of the injection of MSC-EVs that were harvested from the same parent cells. Moreover, the MSC-NVs led to enhanced cardiomyocyte proliferation and promoted angiogenesis in the postinjury heart. Therefore, the extrusion-based technique is an efficient and promising approach to generating relatively massive therapeutic nanovesicles, providing an alternative to EV-like regenerative medicine development [53].

Endothelial dysfunction is the cause of impaired angiogenesis in patients with diabetes, leading to a delayed wound-healing process [54]. As exosomes and EVs emerge as potential carriers for delivering drugs to the diseased cells, Fu et al. reported a novel EV-based therapeutic in treating diabetic wounds by delivery of VH298 into endothelial cells [55]. Epidermal stem cells (ESC)-derived EVs were loaded with VH298 (VH-EVs) via incubation, and the physical and chemical properties of purified VH-EVs were characterized. *In vitro* experiments indicated that the VH-EVs promoted the function of human umbilical vein endothelial cells (HUVECs) via activation of the HIF-1α signaling pathway. In addition, the authors demonstrated that the VH-EVs indicate remarkable therapeutic effects on wound healing and angiogenesis in animal studies. Furthermore, they prepared gelatin methacrylate (GelMA)-based hydrogel with excellent biocompatibility and appropriate mechanical properties to realize the sustained release of VH-EVs. The VH-EVs-containing GelMA hydrogel (Gel-VH-EVs) effectively promoted wound healing via local enhancement of blood supply and angiogenesis process in diabetic mice. The mechanism of enhanced angiogenesis might be related to the activation of HIF-1α/VEGFA signal pathway. In general, this work proposed a promising EV-based strategy for delivering VH298 to endothelial cells, providing a novel bioactive dressing for diabetic wound treatment [55].

Spinal cord injury (SCI) refers to the injury of the spinal cord that changes its function temporarily or permanently. According to the statistical data, about 3 million people are suffering from traumatic spinal cord injury in the world, and there are 180,000 new cases reported each year [56]. SCI usually leads to devastating long-term neurological impairments that are characterized by motor, sensory, or autonomic nervous system dysfunctions below the injury level. The initial mechanical injury damages neurons and non-neuronal cells, destroying spinal cord vessels and the blood–spinal cord barrier, followed by a series of secondary events that further damage the spinal cord tissue so that the microenvironment does not allow regeneration [57]. EVs provide an attractive strategy for developing new therapies for SCI treatment; nevertheless, the progress in this topic is restricted by the following issues: (i) the quality of EV varies with the type and physiological state of the parent cells; (ii) lack of scale-up production and isolation of EVs;

and (iii) difficulties in the development of potency assays. To solve these problems, Kim et al. Preselected preparations of human mesenchymal stem/stromal cells (MSC) with a biomarker that determines their efficacy in regulating anti-inflammatory activities and incubated cells in a chemically-defined medium with no proteins that can provide a stabilized environment [58]. After that, the authors isolated EVs via a scale-up chromatography protocol. Furthermore, they tested the efficacy of the isolated EVs in regulating inflammation in a mouse model of traumatic brain injury. The results indicated that the intravenous injection of isolated EVs after TBI induction saved pattern separation and spatial learning disabilities in mice [58].

Dental pulp is a kind of soft tissue that is rich in cells, blood vessels, and nerves, and is essential for dental nutrition, sensation, and dentin formation [59]. Dental pulp infection is usually caused by severe dental caries and injury, leading to pulpitis and pulp necrosis, pulp removal, and even tooth loss. Root canal therapy (RCT) is a classical treatment strategy for dental pulp-related diseases. Nevertheless, RCT could not restore or promote the regeneration of dental pulp tissue, especially neurogenesis and neovascularization [60]. Schwann cells (SCs) are reported to play a significant role in the support, maintenance, and regeneration of dental pulp nerve fibers. Stem cell-derived EVs indicated the properties of cell homing and tissue repair. SC-derived EVs (SC-EVs) showed their functions in regulating the proliferation, pluripotency, and self-renewal of dentin MSCs [61]. Nevertheless, their role in dental pulp tissue regeneration is still unclear. In order to figure out the mystery, Wang et al. used SC-EVs to treat dental pulp stem cells (DPSCs) and bone marrow stem cells (BMSCs) (Figure 4.4) [62]. The results demonstrated the increased proliferation, migration, and osteogenic differentiation of these two cell lines after SC-EV treatment. In addition, *in vitro* experiments indicated that SC-EVs promoted neurite outgrowth, neuron migration of dorsal root ganglion, and angiogenesis. *In vivo* hypodermic model, Also, SC-EVs significantly recruited endogenous vascular endothelial-like cells and MSCs and enhanced the pulp–dentin complex-like structure formation in an *in vivo* model. In the end, the authors conducted mass spectrometry and western blotting analysis, and figured out that stromal cell-derived factor 1 (SDF-1) played an important role in SC-EVs. Altogether, this study demonstrated the function of SC-EVs in the recruitment of endogenous stem cells to promote dental pulp regeneration, providing a cell-free strategy for pulp regeneration [62].

The periodontal disease begins with an inflammatory reaction to the bacterial biofilm deposited around the teeth, which gradually leads to the destruction of the supporting structure of the teeth and eventually to tooth loss [63]. Traditional treatment strategies have shown limited effectiveness in promoting the regeneration of damaged periodontal tissue. EVs could achieve a similar therapeutic effect as their source cells. However, the short half-life of EVs after *in vivo* administration and their rapid diffusion from the delivery site demands multiple administrations to realize the satisfactory therapeutic outcome [64]. In order to solve these problems, Zarubova et al. proposed a delivery platform for small extracellular vesicles (sEVs) from gingival mesenchymal stem cells (GMSCs) to treat periodontitis (Figure 4.5) [65]. The authors designed micro-sized "dual delivery" particles to release sEVs in

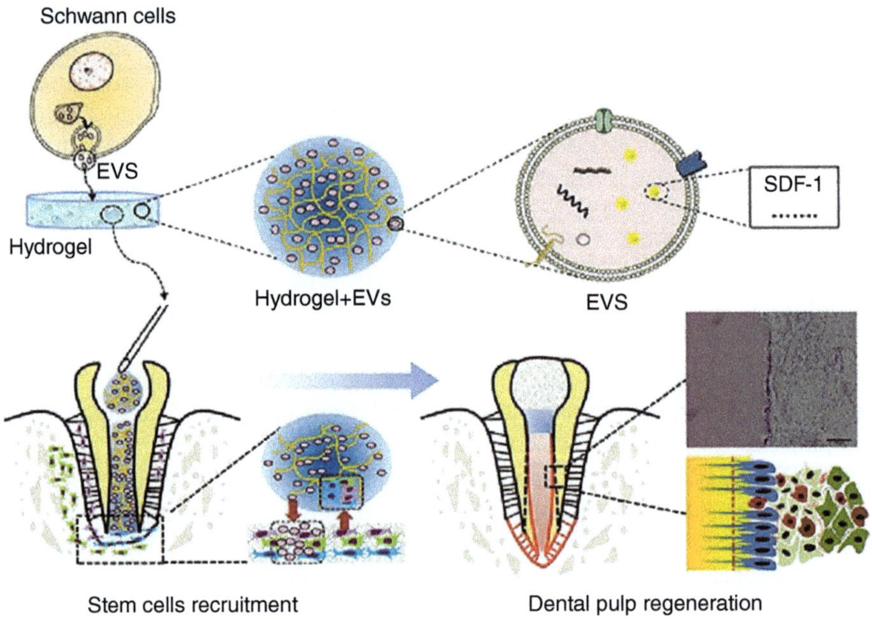

Figure 4.4 Scheme of Schwann cells-derived EVs released from hydrogel for dental pulp regeneration. Source: Wang et al. [62]/with permission of Elsevier.

response to the microenvironment through metalloproteinases at the disease site, and antibiotics were used to inhibit the growth of bacterial biofilm [66]. GMSC sEVs were selected for micromodification, as these nanoparticles indicated better immunomodulatory properties than sEVs derived from other types of mesenchymal stem cells [67]. In their design, sEVs were fixed to microparticles through MMP2-sensitive connectors, enabling sEVs to specifically locate at the site of tissue injury and prolong their presentation, which leads to enhanced periodontal tissue regeneration. The results indicated that the GMSC sEVs can reduce the secretion of proinflammatory cytokines in monocytes, macrophages, and T cells. In addition, the treatment of GMSC sEVs efficiently inhibited T cell activation and induced Treg cell formation *in vitro* and an animal model of periodontal disease. The single administration of microparticles modified with immunomodulatory GMSC sEVs led to significant improvement in the regeneration of damaged periodontal tissue. This method has potential clinical application in many kinds of tissue regeneration [65].

Though MSC-EVs could lead to rapid cell proliferation and migration with no obvious immune responses and indicated promising potential in wound healing, their application in tissue injuries that happen in our daily life is still challenging in clinical medicine [68]. To this end, the development of novel biomaterials that provide a suitable microenvironment and balance at the wound site is beneficial for displaying the full potency of tissue regenerative medicine. The previous studies indicated that the small intestinal submucosa (SIS) membrane has precise spatial architecture and good biocompatibility [69]. Accordingly, Zhang et al. modified the SIS membrane surface via a specific combination of umbilical cord mesenchymal

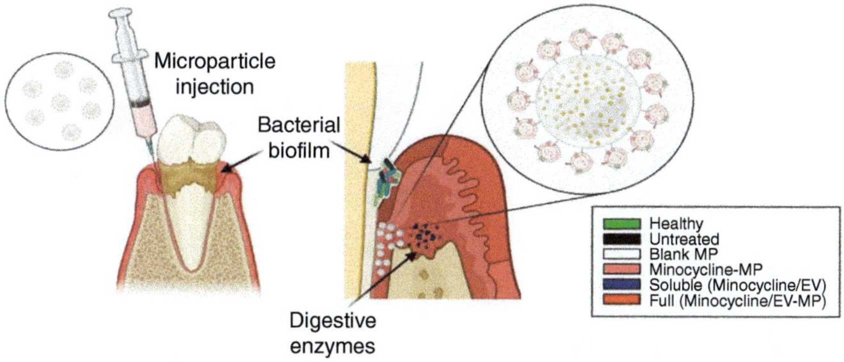

Figure 4.5 Microparticles containing gingival mesenchymal stem cells-derived EVs for regeneration of damaged periodontal tissue. Source: Zarubova et al. [65]/John Wiley & Sons.

stem cell-derived EVs (UMSC-EVs) using fusion peptide (Figure 4.6) [70]. In detail, CP05 is a small molecular peptide that can specifically bind to CD63—a protein that is enriched on the surface of EVs [71]. As the main components of the SIS membrane are type I and type III collagens, two collagen-binding peptides—LHERHLNNN and KELNLVY that can specifically anchor biomolecules on the surface of type I and III collagens were selected in this work. Accordingly, the authors synthesized a series of fusion peptides to specifically bind and load LHERHLNNN and KELNLVY on the SIS membrane via conjugation of EVs and CP05 with or without a flexible GGGGS connector. Such modification provided the SIS membrane with a better therapeutic effect of tissue regeneration. *In vitro* experiments demonstrated that the modified SIS membrane can significantly promote cell migration and diffusion. This phenomenon might be due to the activation of transcriptional-enhanced associate domains (TEADs) that regulate cell activities. The authors further indicated that the modified SIS membrane is more beneficial to tissue regeneration in a rat abdominal wall defect model. In general, this work showed that the SIS membrane with fusion peptide-mediated EV modification possesses remarkable biological functions and provides broad prospects for tissue regeneration [70].

4.2 Metabolic Diseases

Metabolic diseases refer to any types of diseases or disorders that disrupt the normal cellular metabolism process. Metabolic diseases are usually caused by the dysfunctional crosstalk among various cells, such as adipocytes, hepatocytes, and immune cells [72]. In typical, the cellular crosstalk is medicated by hormones and metabolites, changes of which result in metabolic syndrome, such as obesity and diabetes. As the basic function of EVs is transmitting information through the delivery of proteins and nucleic acids between cells and organs, EVs represent an attractive strategy for treating metabolic diseases via mediating intercellular communication.

Diabetes, recognized as a global metabolic disease, is characterized by insufficient insulin release from pancreatic cells and elevated blood glucose levels [73]. To

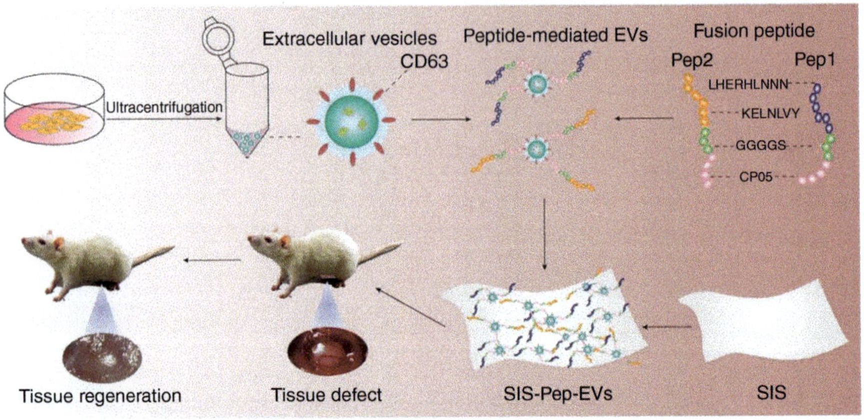

Figure 4.6 Schematic demonstration of small intestinal submucosa membrane functionalized with umbilical cord mesenchymal stem cell-derived EVs (UMSC-EVs) for Tissue Regeneration. Source: Zhang et al. [70]/with permission of John Wiley & Sons, Inc.

maintain blood glucose at a normal level, patients need to undergo continuous blood glucose monitoring and frequent insulin injections, bringing a great inconvenience to patients [74]. In addition, this open-loop routine insulin administration sometimes is accompanied by hypoglycemia, leading to serious side effects, such as blindness, kidney failure, and even death [75]. To reduce the risk of hypoglycemia, a closed-loop insulin delivery system that can rapidly respond to hyperglycemia and maintain basal insulin release at normal blood glucose levels is more advantageous. Milk-derived exosomes indicate similar potential as drug delivery nanoparticles as a more affordable and accessible source than cell culture [76]. Accordingly, Wu et al. fabricated insulin-loaded milk-derived exosomes (EXO@INS) and investigated the hypoglycemic effect on type I diabetes (Figure 4.7) [77]. The results indicated that oral administration of EXO@INS had a superior and more sustained hypoglycemic effect compared to subcutaneous injection of insulin. After that, the authors studied the mechanisms and demonstrated that the oral delivery of EXO@INS leads to active multi-targeting uptake, pH adaptation at the gastrointestinal transit process, activation of ERK1/2 and p38 MAPK signal pathways related to nutrient assimilation, and penetration into intestinal mucus. This work provided novel insights into the development of natural nanovesicles for oral drug delivery for diabetes treatment [77].

MSCs themselves have also been proved to reduce diabetes mellitus (DM) in animal models and clinical trials [78]. For example, Sun et al. investigated whether the exosomes of human umbilical cord mesenchymal stem cells (hucMSC-ex) possess therapeutic effects on T2DM (Figure 4.8) [41]. Type 2 diabetes mellitus (T2DM) accounts for 95% of diabetes cases. Recently, T2DM has been called an inflammatory state caused by metabolic disorders [79]. The authors established the T2DM rat model with a high-fat diet (HFD) and streptozotocin (STZ). After that, the authors figured out that the intravenous injection of hucMSC-ex significantly reduced blood glucose levels mainly due to the result of the paracrine approach

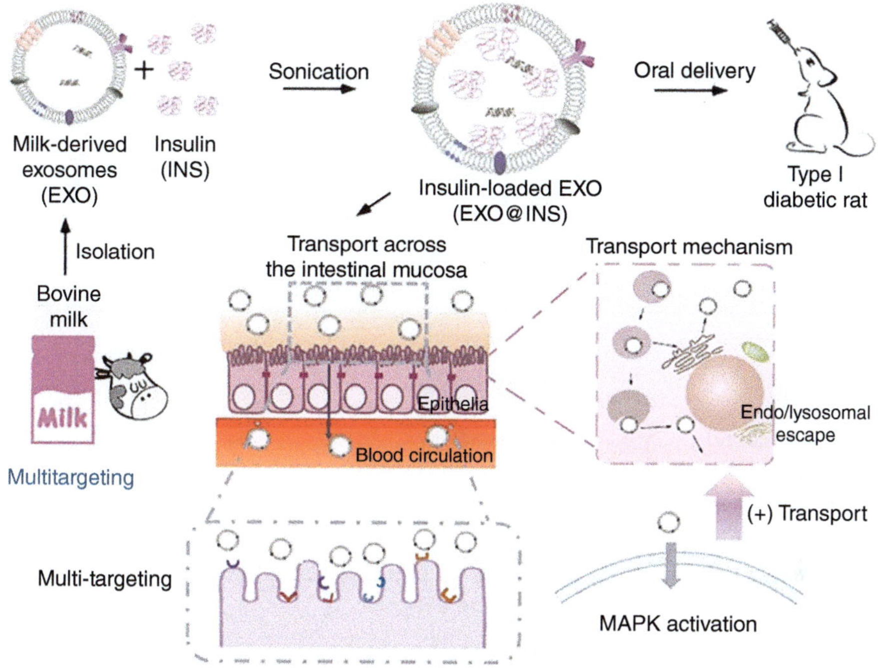

Figure 4.7 Oral administration of insulin-loaded milk-derived exosomes indicated a remarkable hypoglycemic effect on type I diabetes. Source: Wu et al. [77]/with permission of Elsevier.

of MSC. HucMSC-ex was proved to partially reverse the insulin resistance in T2DM and accelerate glucose metabolism. In addition, they demonstrated that the hucMSC-ex restored the tyrosine phosphorylation of insulin receptor substrate 1 (IRS-1) and protein kinase B in T2DM, leading to the promoted expression and membrane translocation of glucose transporter 4 (GLUT4) in muscle, increased glycogen storage in liver, and maintained glucose homeostasis. Also, hucMSC-ex could restore the insulin secretion function in T2DM via inhibition of apoptosis of β cells induced by STZ. In summary, this work confirmed that hucMSC-derived exosomes can mitigate T2DM via reversing peripheral insulin resistance and promoting β-cell survival, providing an attractive alternative in T2DM therapy [41].

In another study, Zeinab et al. investigated the effect of salivary exosomes (salivary-Exos) on the improvement of T2DM in diabetic rats and the treatment of xerostomia which is a complication of submandibular gland dysfunction [80]. The rats were randomly distributed into two groups: Group I (diabetes group) without any treatment, and Group II (Salivary Exo-treated group) with intravenous injection of salivary-Exos one week after induction of diabetes. The blood glucose level of rats in these two groups was determined every week. In addition, the water intake, salivary flow rate and amylase, and nitric oxide in the serum were measured before and after diabetes induction, as well as at the end of the study. After five weeks, the salivary glands were harvested and examined. The expression of NFκB/p65

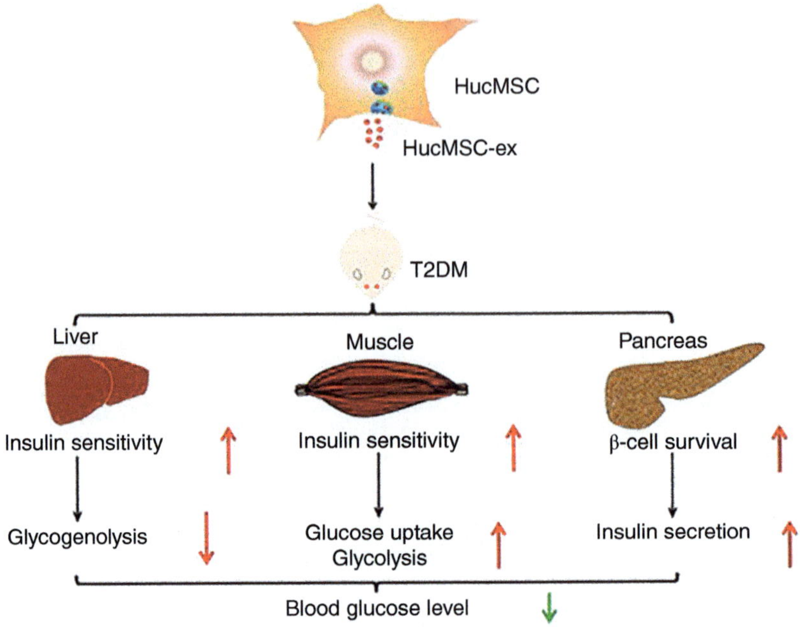

Figure 4.8 Exosomes derived from human umbilical cord mesenchymal stem cells (hucMSC-ex) mitigate T2DM via reversing peripheral insulin resistance and reducing β-cell apoptosis. Source: Adapted from Qi et al. [41].

and TNF-α was also studied via PCR. The experimental data indicated that the salivary-Exos significantly decreased the level of blood glucose and enhanced the functions of the salivary gland. This statement is proved by the reduced water intake, salivary amylase and serum nitric oxide, increased salivary flow rate, and downregulated expressions of NFκB/p65 and TNF-α. This work suggested that salivary-Exos is a novel potential cell-free therapy for the treatment of xerostomia and salivary gland dysfunction in DM [80].

The decline of β-cell quality in T2DM is attributed to the suppressed number and functions of β-cells [81]. Recently, it was suggested that the loss of functional β cells is due to dedifferentiation instead of apoptosis [82]. The dedifferentiated β cells lose the capability of insulin secretion and transform into endocrine progenitor cells via the activation of angiotensin II receptor type 1 promoted by the NF-κb signaling pathway [83]. Blocking the β-cell dedifferentiation process could reverse the hyperglycemia phenotype and slow down the diabetes progression. Therefore, safe and effective treatment for reversing β-cell dedifferentiation may represent a novel and promising treatment strategy for T2DM. He et al. clarified the therapeutic effect of bone marrow MSC (bmMSCs) on β-cell dedifferentiation (Figure 4.9) [84]. Both *in vitro* and *in vivo* results indicated that bmMSC-derived exosomes (bmMDEs) could reverse diabetic β-cell dedifferentiation and improve the insulin secretion function of β-cell. miRNA sequencing identified the highly conserved miR-146a as the mediator of the beneficial effect of bmMDEs on β-cell diabetic dedifferentiation and function. The

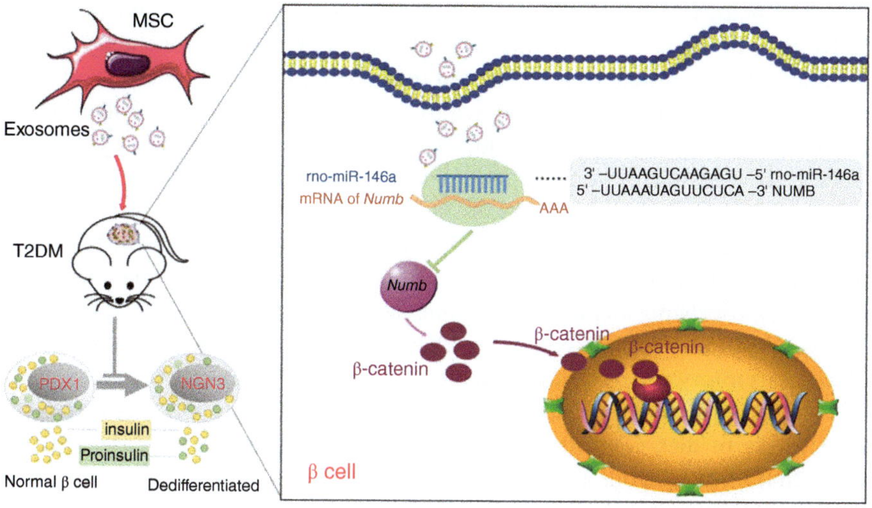

Figure 4.9 Schematic illumination of reverse of diabetic β-cell dedifferentiation by bmMSC-derived exosomes via the miR-146a/*Numb*/β-catenin signaling pathway. Source: He et al. [84]/Springer Nature/CC BY 4.0.

authors further demonstrated that miR-146a can directly target Numb, which is a membrane-binding protein related to cell fate determination, subsequently activating the β-catenin signaling pathway in β-cells. This work illuminated the therapeutic potential of bmMSC-derived exosome-based therapy in treating diabetes [84].

It was also reported that diabetes (DM) increases the risk of fracture from 20% to 300% [85]. In bone tissue engineering, the combination of biomimetic coatings and bioactive materials in a scaffold is a common strategy for functional enhancement. Inorganic materials, such as phosphate and silicate materials [86], have been widely tested in bone tissue engineering. Nevertheless, the single introduction of such materials in large bone defect areas is not sufficient, especially when in combination with uncurable diseases, for example, DM and osteoporosis [87]. Such diseases remarkably extend the healing process of bones. Therefore, reasonable optimization of inorganic materials to enhance their biological activity is an issue worth solving. To this end, Tao et al. combined commercial porous β-TCP (β-tricalcium phosphate) scaffold with modularized engineered sEVs for the regeneration of bone defects in DM condition [88]. In detail, the coating of hyaluronic acid (HA)/poly-L-lysine (PLL) was developed on β-TCP via a layer-by-layer (LbL) self-assembly technique. After that, sEVs were engineered with surface functionalization of DSPE-PEG-c(RGDfC) and zinc finger e-box binding homeobox 1 (ZEB1) loading via a system called "exosomes for protein loading via optically reversible protein-protein interactions" (EXPLOR). Then, the modularized engineered sEVs were fixed on HA/PLL coating. Synovial MSCs (SMSCs) that can produce two kinds of recombinant fusion proteins—CRY2-ZEB1 and CIBN-CD9, which are linked by a flexible linker (3 × GGGGS), were chosen for the production of ZEB1-loading sEVs via transient docking of CIBN and CRY2 in a blue-light manipulative manner.

RNA sequencing (RNA-Seq) technique was utilized to study the possible molecular mechanism. In addition, the therapeutic efficacy in promoting bone defect regeneration of DM was systematically assessed *in vivo* and *in vitro*. The results indicated that the proposed strategy was effective in enhancing angiogenesis, osteogenesis, and inhibiting osteoclast formation during the repair of diabetic bone defects [88].

EVs, which function as information transferring substances, could deliver nucleic acids and peptides to the recipient cells and mediate the paracrine effects of parent cells [89]. The previous studies indicated that intrarenal delivery of MSC-EVs could retain the stenotic renal function and reduce proinflammatory cytokine release in a pig model of coexisting metabolic syndrome (MetS) and renal artery stenosis (RAS). Accordingly, Zhang et al. hypothesized that this technique can also be utilized to reduce heart injury and dysfunction in metabolic renovascular disease (Figure 4.10) [90]. Five groups were designed, including MetS and RAS (MetS+RAS) induced by 16 weeks of diet, MetS treated with intrarenal delivery of EVs-rich fraction harvested from autologous adipose tissue-derived MSCs 4 weeks earlier, MetS+RAS treated with intrarenal delivery of EVs-rich fraction harvested from autologous adipose tissue-derived MSCs 4 weeks earlier, lean, and MetS Shams. *In vivo*, cardiac structure, function, and myocardial oxygenation and ex vivo cardiac inflammation, senescence, and fibrosis were studied using imaging techniques. Also, the inflammatory cytokine levels in circulating blood and renal vein blood were determined. The results indicated that the intrarenal delivery of MSC-EVs greatly promoted stenotic-kidney glomerular filtration and blood flow in the kidney, and reduced the release of monocyte chemoattractant protein-1 and interleukin-6 in the kidney. In addition, intrarenal EV delivery in the MetS+RAS group normalized cardiac diastolic function, reduced remodeling of the left ventricle, cell senescence, and inflammation, as well as improved myocardial oxygenation and capillary density, though there was no significant change in systemic hemodynamics. The attenuation of myocardial injury in experimental MetS+RAS by intrarenal delivery of MSC-EVs might be related to the improvement of renal function and systemic inflammation activities. This work highlighted the role of inflammation in crosstalk between the kidney and the heart and the contribution of renal function in maintaining the structural and functional integrity of the heart in coexisting MetS and RAS [90].

The inherent regeneration ability of articular cartilage after the injury is limited. If cartilage injury is not properly treated, it can lead to osteoarthritis (OA)—an inflammatory degenerative joint disease that affects the entire joint, leading to pain, deformity, and loss of function [91]. With the aging of the population, the incidence of OA is also increasing. The incidence of OA in middle-aged people is as high as 40%–80%, and the disability rate is more than 50%, posing a huge burden on individuals and society [92]. At present, the treatment of OA is mainly focused on the development of drugs to relieve pain and control inflammation, but these treatments cannot prevent or treat cartilage destruction in OA [93]. Many studies have focused on sEVs and tried to determine their potential biomechanics in the treatment of various diseases [94]. Hypoxic preconditioning could improve the therapeutic effects of MSCs and regulate the expression of specific miRNAs. Nevertheless, the functions of miRNAs in sEVs derived from MSCs in hypoxia conditions and their potential

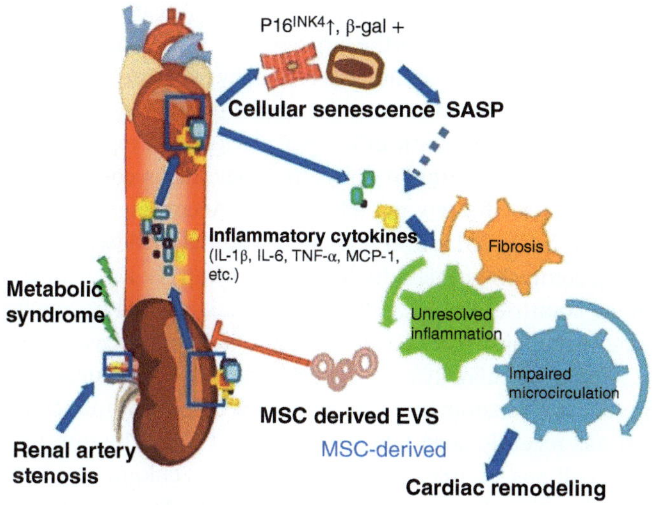

Figure 4.10 Intrarenal delivery of MSC-derived EVs for myocardial injury attenuation in a model of coexisting metabolic syndrome and renal artery stenosis. Source: Zhang et al. [90]/Springer Nature.

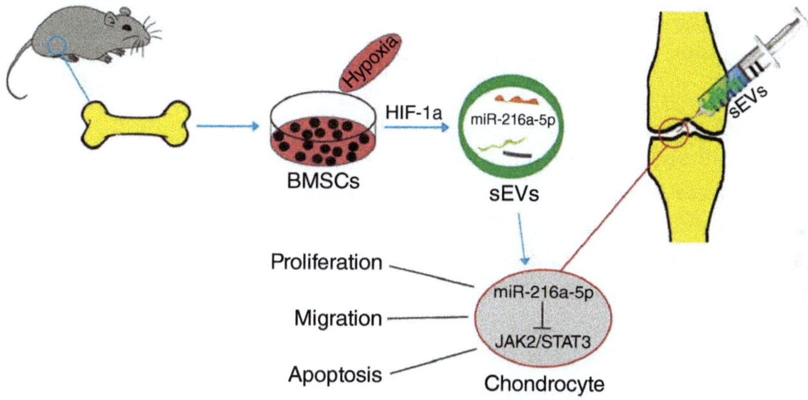

Figure 4.11 Schematic demonstration of hypoxia-pretreated small EVs (sEVs) for cartilage repair in osteoarthritis via the delivery of miR-216a-5p. Source: Rong et al. [95]/with permission of Elsevier.

biomechanical mechanism of promoting OA repair *in vivo* are not investigated. Accordingly, Rong et al. verified the effect of hypoxic preconditioning on the biological activity of MSC-sEVs regarding the regulation of miRNA expression and OA repair by comparing the sEVs in hypoxia (Hypo-sEVs) state and sEVs in normoxia (Nor-sEVs) state (Figure 4.11) [95]. The results showed that Hypo-sEVs treatment had a stronger effect in enhancing the proliferation, migration, and suppressing the apoptosis of chondrocytes than Nor-sEVs. Using a miRNA chip, the authors determined that the miR-216a-5p was enriched in hypo-sEVs. They further confirmed that miR-216a-5p silencing in Hypo-SEVs eliminated the beneficial effects, proving the central role of Hypo-SEVs-derived miR-216a-5p in the OA cartilage repair process. After that, they identified that *JAK2* is the target gene of Hypo-SEVs-derived miR-216a-5p by bioinformatics analysis and molecular biology experiments. Therefore, this work suggested that hypoxic pretreatment might be a prospective and effective approach to optimize the therapeutic efficacy of sEVs on OA repair [95].

Obesity, defined as excessive fat accumulation in the body fat, is considered a global public health challenge for children and adolescents. According to a report of the World Health Organization in 2016, over 650 million adults worldwide suffer from obesity. It used to be a common problem in developed countries; nevertheless, it also occurs in developing countries nowadays [96]. Usually, increased body mass index (BMI) is considered the major risk factor for obesity-related diseases, such as metabolic syndrome, cardiovascular disease, type 2 diabetes, and cancer [97]. Obesity is usually a multifactor phenomenon, in which the intestinal microbiota is one of the factors that has recently been taken into account [98]. The intestinal microbiota is defined as trillions of microorganisms that are colonized in the gastrointestinal tract and play a vital role in human health [99]. After the change in diet, the intestinal microbiota changes immediately, affecting many molecular pathways. In recent years, studies have shown that probiotics could affect the composition of intestinal microorganisms, restore the integrity of the mucosal barrier, improve inflammatory responses, and promote the dynamic metabolism balance in patients with obesity. Accordingly, Ashrafian et al. studied the influence of *Akkermansia muciniphila* (*A. muciniphila*) and derived EVs on obesity-related genes *in vitro* and *in vivo* (Figure 4.12) [100]. A meta-analysis of obesity-related genes was performed on nine high-fat diet (HFD) gene chip data sets. The authors identified several adiposity and inflammatory-related genes that were expressed at higher levels in the obese compared to normal and selected these genes to investigate the effect of *A. muciniphila* and EVs. The *A. muciniphila*-derived EVs induced significant loss of both body and fat weight in HFD-fed mice. In addition, administration of *A. muciniphila* and EVs showed obvious effects on lipid metabolism and inflammatory marker expression in epididymal adipose tissue (EAT). Both *A. muciniphila* and EVs significantly enhanced the structural integrity of the intestinal barrier, inflammation, energy balance, and blood parameters. Their data indicated the positive impact of *A. muciniphila*-derived EVs on obesity via regulation of involved genes, suggesting that *A. muciniphila*-derived EVs could be used as a novel therapeutic strategy for the amelioration of HFD-induced obesity [100].

4.3 Cardiovascular Diseases

Cardiovascular diseases (CVDs), including coronary heart disease (CHD), cerebrovascular disease, hypertension, and peripheral vascular disease (PVD), are a broad group of diseases affecting the heart and blood vessels [14, 101]. CVDs are also known as circulatory system diseases, referring to ischemic or hemorrhagic diseases of the heart, brain, and body tissues caused by hyperlipidemia, blood viscosity, atherosclerosis, hypertension, etc. [102–104]. Extracellular vesicles (EVs),

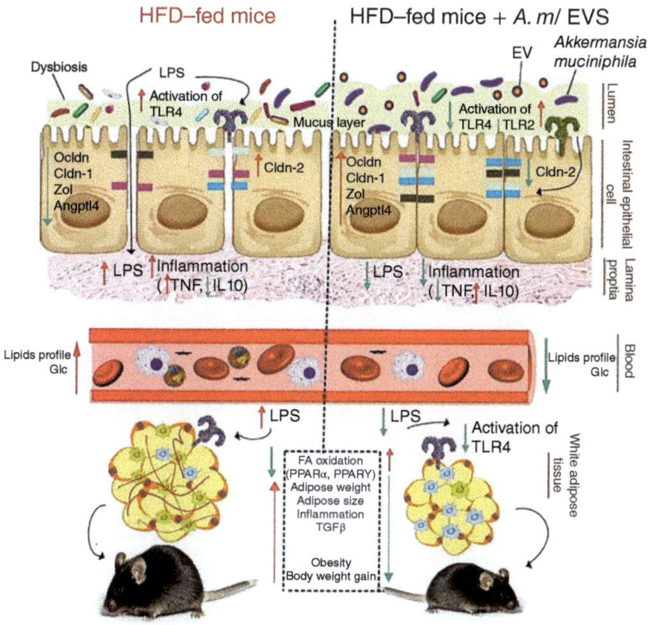

Figure 4.12 Treatment of *Akkermansia muciniphila* and derived EVs promotes intestinal and metabolic homeostasis in mice with obesity. Source: Ashrafian et al. [100]/Frontiers Media S.A./CC BY 4.0.

because of their unique vesicle structure, can not only protect the internal carrier substances from degradation but also rapidly transport the cargoes of interest into recipient cells without inducing cytotoxicity and adverse immune reactions.

Myocardial infarction (MI) refers to acute myocardial ischemia necrosis, which is mostly caused by severe and lasting acute ischemia of the corresponding myocardium caused by sharp reduction or interruption of coronary artery blood supply based on coronary artery lesions. It belongs to acute coronary syndrome [105, 106]. An increasing number of studies show that RNAs are involved in the regulation of post-MI cardiac angiogenesis, apoptosis, and fibrosis, making them potential pharmacological candidates for MI treatment [107–109]. Thus, recent research has considered the use of EVs as RNA carriers. As a novel RNA delivery vesicle, EVs demonstrated high miRNA loading efficacy, excellent biocompatibility, and high cellular uptake efficiency bypassing the endosomal obstacle [35]. Accordingly, Ge et al. described a "two-step" strategy toward modifying MSC-EVs and demonstrated the therapeutic effects of miRNAs (Figure 4.13) [110]. Firstly, they performed bioinformatic analysis to highlight key molecules miRNA21 within MSC EVs and increase the content level of certain cargo by electroporation. Secondly, they constructed MSC-EVs overexpressing CD47 to reduce recognition and phagocytosis in the circulation and prolong the retention time of EVs *in vivo*. Here, they provide a novel potential strategy to prolong the EV retention time in circulation. In conclusion, electro genetically modified electro CD47-EVs can successfully accumulate in

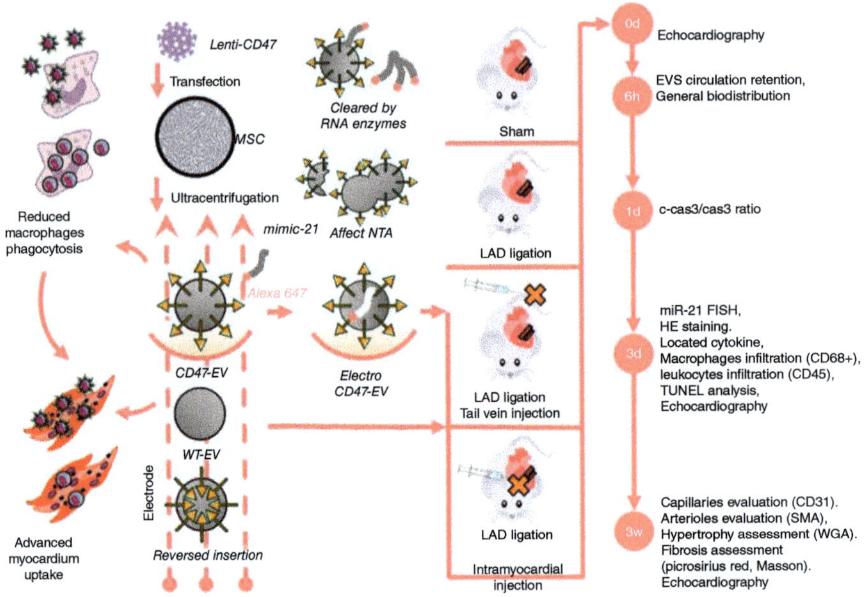

Figure 4.13 EVs with surface modification with CD47 for treatment of myocardial infarction reperfusion injury mediated by miRNA21. Source: Wei et al. [110]/with permission of Elsevier.

the heart and promote the therapeutic effect on myocardial infraction/reperfusion injury., with prolonged retention in circulation [110].

In another study, human pluripotent stem cells (hPSCs)-derived cardiovascular progenitor cells (CVPCs) have shown a potential effect to improve cardiac function after myocardial infarction in rodents [111]. Yang et al. isolated hPSC-CVPC-secreted EVs (hCVPC-EVs) and found that the infusion of hCVPC-EVs into the early phase of acute myocardial infarction (AMI) can promote recovery of damaged cardiac function and reduce the occurrence of myocardial fibrosis (Figure 4.14) [112]. Moreover, the hCVPC-EVs contain lncRNA MALAT1, which partially contributed to the inhibition of cardiomyocyte death and the promotion of angiogenesis via targeting miR-497. In addition, the MALAT1 abundance in the hCVPC-EVs can be upregulated by hypoxia conditioning of hCVPCs. These findings provided new insights into the cardioprotective mechanisms of hCVPCs and the cell-secreted EVs and suggested that hCVPC-EVs might be used as a tool to understand the mechanism of infarct healing and promote healing of infarcted hearts [112].

Very recently, Wang et al. first synthesized a type of circular RNA (CircUbe3a) and loaded it into small EVs (SEVs) derived from M2 macrophage (M2M) for the treatment of myocardial fibrosis after MI [113]. The authors found that M2M is involved in mediating cardiac fibrosis after MI by releasing SEVs containing circUbea3a and regulates the expression of *Rho*C, which is the target gene of miR-138-5p and can be reversed by miR-138-5p. Further analysis demonstrated that

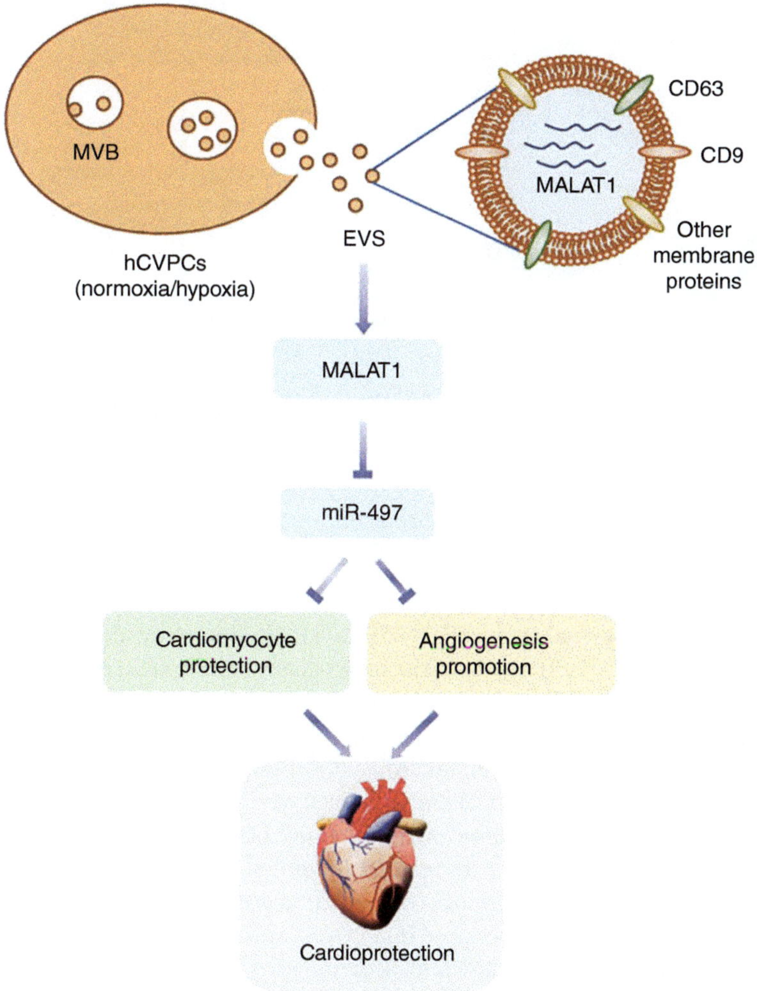

Figure 4.14 Schematic demonstration of therapeutic effect of human pluripotent stem cells -secreted EVs (hCVPC-EVs) in the infarcted heart. Source: Wu et al. [112]/Springer Nature/CC BY 4.0.

the circUbe3a played a significant role in promoting the proliferation, migration, and myoblast transformation of cardiac fibroblasts (CFs) by transferring circUbe3a to receptor cells, targeting the circUbe3a/miR-138-5p/RhoC axis. It is shown that SEV-mediated intercellular communication between M2Ms and CFs may lead to pathogenic ventricular remodeling after acute myocardial infarction (AMI), which may also exacerbate myocardial fibrosis after AMI [113].

In addition, it was reported that stem cell-derived sEVs promotes angiogenesis after MI [114]. For example, Hu et al. observed that sEVs from hypoxia preconditioning (HP) mesenchymal stem cells (MSCs) (HP-sEVs) could improve cardiac function, increase vascular density, and lead to smaller infarct size than normal-oxygen

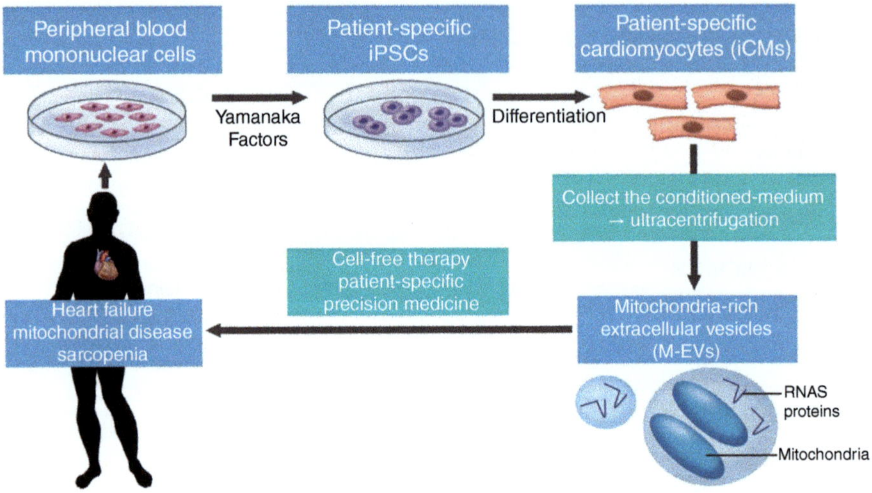

Figure 4.15 Schematic illumination of mitochondria-rich EVs derived from iPSCs-derived cardiomyocytes for treatment of heart failure. Source: Ikeda et al. [119]/with permission of Elsevier.

preconditioning mesenchymal stem cells (N-SEVs) [115]. This study identified the metalloproteinase 19 (MMP19) as one of the target genes of miR-486-5p and highlighted the key role of sEVs miR-486-5p in promoting cardiac angiogenesis via fibroblastic MMP19-VEGFA cleavage signaling pathway. It also showed that delivery of miR-486-5p-engineered sEVs safely enhanced angiogenesis and cardiac function in a nonhuman primate (NHP) MI model and promoted cardiac repair [115].

In another study, Xu et al. focused on the therapy of myocardial ischaemia–reperfusion (I/R) injury after acute myocardial infarction [116]. They found that administration of MSC-derived exosomes (MSC-Exo) to mice through intramyocardial injection after myocardial I/R reduced infarct size, and alleviated inflammation levels in the heart and serum. Meanwhile, accumulating evidence indicated that MSC-Exo can be able to mediate the polarization of M1 macrophages to M2 macrophages [117]. The authors also confirmed this statement and demonstrated that miR-182 was abundant in MSC-Exo and regulated macrophage phenotype. They examined two signaling pathways and found that PI3K/AKT were activated and TLR4/NF-κB was inhibited after MSC-Exo treatment or miR-182 transfection. The inhibition of the TLR4/NF-κB pathway and activation of the PI3K/Akt pathway was partially reversed when miR-182 was silenced in exosomes. These results indicated that MSC-Exo could attenuate myocardial I/R injury in mice via miR-182 to modify the polarization status of macrophages by targeting the TLR4/NF-κB and PI3K/Akt signaling pathways [116].

It is reported that EVs could facilitate the secure and efficient transfer of their mitochondrial cargo into the recipient cells [118]. Accordingly, Yang et al. developed human-induced pluripotent stem cells (iPSCs)-derived cardiomyocytes (iCMs) and succeeded in collecting mitochondria-rich EVs (M-EVs) from the iCM-conditioned medium (Figure 4.15) [119]. Intramyocardial injection of M-EVs into the peri-infarct

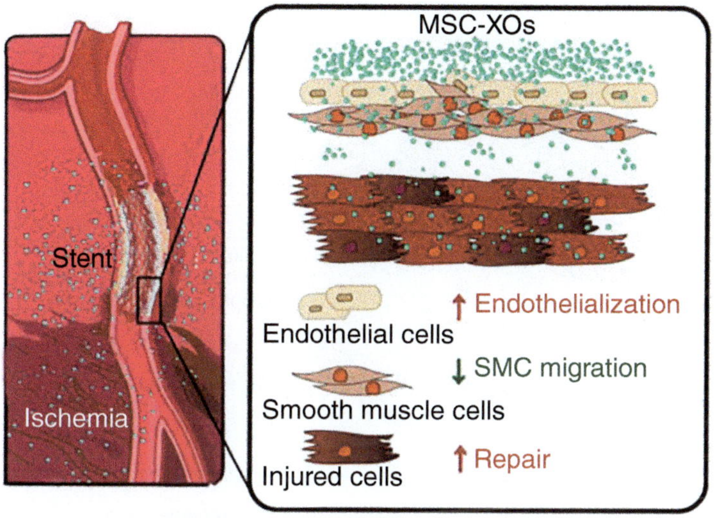

Figure 4.16 Schematic demonstration of exosome-eluting stents for the treatment of ischemic tissue. Source: Hu et al. [120]/Springer Nature.

region significantly prevented post-MI left ventricular remodeling as compared with the control group in a mouse model of MI. The protein expressions of PGC-1a and COX IV were upregulated at day 28 after MI. It suggested that the M-EV therapy could restore bioenergetics and facilitate mitochondrial biogenesis in the peri-infarct region. This demonstrated the feasibility of M-EV-mediated transfer of mitochondria and their related bioenergetics cargo for the treatment of failing hearts. Altogether, this work provided a novel and effective therapy for mitochondria-related diseases, including advanced heart failure patients [119].

Cheng et al. also developed several EV-based strategies for the treatment of cardiovascular diseases. For example, Hu et al. developed a stent for sustained release of MSCs-derived exosomes (MSC-Xos) that could not only regulate vascular remodeling and inflammation but also promote the regeneration of the injured tissue, taking advantage of the elevated level of reactive oxygen species (ROS) from the mechanical injury (Figure 4.16) [120]. The results demonstrated that vascular stents can be ideal carriers to deliver therapeutic exosomes to the ischemic tissue. Synergistically, the exosome coating on stents could improve biocompatibility, inhibit in-stent restenosis (ISR), and promote vascular healing. Compared with drug-eluting stents and bare metal stents, the exosome-coated stents accelerated reendothelialization and decreased in-stent restenosis 28 days after implantation. The data also showed that exosome-eluting stents implanted in the abdominal aorta of rats with unilateral hindlimb ischemia regulated macrophage polarization, reduced local vascular and systemic inflammation, and promoted muscle tissue repair. In addition, exosome-eluting stent promoted endothelial tube formation, and proliferation and inhibited the migration of smooth muscle cells (SMCs) [120].

In another study, Qiao et al. isolated and compared the therapeutic effects of exosomes from explant-derived cardiac stromal cells in patients who have heart

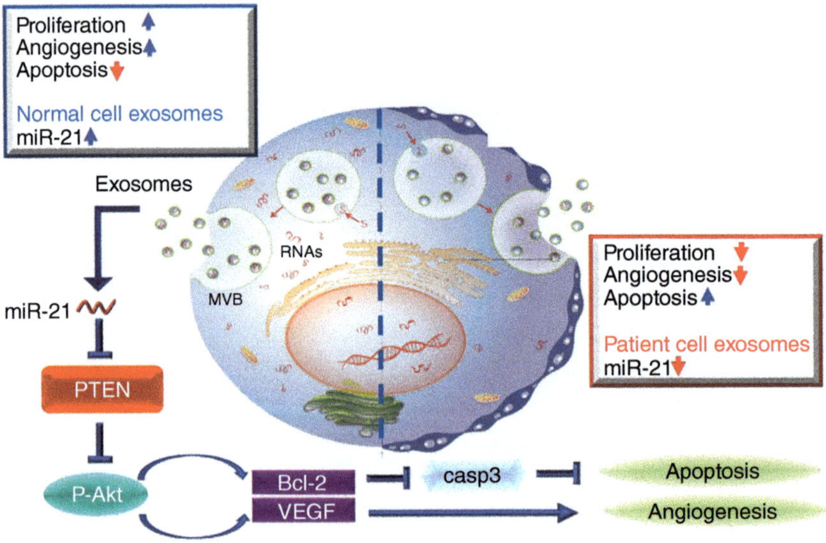

Figure 4.17 Alteration of miRNA expression in cardiac-derived exosomes by heart pathological condition. Source: Qiao et al. [121]/American Society for Clinical Investigation.

failure (FEXO) or normal hearts (NEXO) (Figure 4.17) [121]. The results indicated that the intramyocardial injection of NEXO significantly improved the structural and functional features in a murine model of acute MI. On the contrary, FEXO treatment deteriorated cardiac functions and left ventricular remodeling. The authors conducted a microRNA array and PCR analysis and found the dysregulation of miR-21-5p in FEXO. Recovering the expression of miR-21-5p rescued the reparative functions of FEXO. In addition, silencing the miR-21-5p expression in NEXO reduced their therapeutic effects. Further analysis demonstrated that miR-21-5p could enhance the activity of Akt kinase via the suppression of phosphatase and tensin homolog. This work suggested that the heart's pathological condition could alter the miRNAs in cardiac-derived exosomes and impair the regenerative potentials [121].

Zhu et al. also investigated the immunomodulation properties and mechanisms of MSCs-derived exosomes in cardiac injury in a MI model (Figure 4.18) [122]. The results indicated that the intrapericardial (iPC) administrated MSC-exosomes were absorbed by MHC-II$^+$ antigen-presenting cells (APCs), subsequently inducing activation of Foxo3 through modulation of the PP2A/p-Akt/Foxo3 pathway. In addition, Foxo3 significantly enhanced the secretion of IL-10, IL-33, and IL-34 by APCs to induce regulatory T cells (Treg) in the mediastinal lymph node (MLN). This work illuminated the critical role of Foxo3 in APCs in the immunomodulation properties of MSC-exosomes in inducing Tregs for post-MI repair [122].

To further enhance the retention of exosomes in the heart, Vandergriff et al. developed infarct-targeting exosomes derived from cardiac stem cell (CHP-EXOs) via conjugation of cardiac homing peptide (CHP) through DOPE-NHS linker [123]. The results demonstrated the superior therapeutic effects of CHP-EXOs over

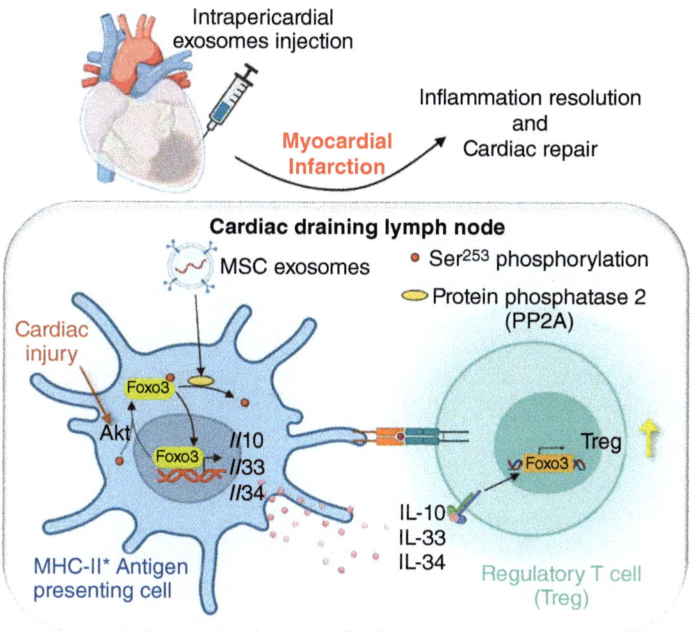

Figure 4.18 Schematic representation of Treg induction by MSC-exosomes via Foxo3 activation. Source: Zhu et al. [122]/Wolters Kluwer Health, Inc.

normal CDCs-exosomes, as CHP-EXOs significantly enhanced the regenerative activities, increased the efficacy and decreased the effective dose. In addition, the imaging data demonstrated that the targeting exosomes led to increased uptake in cardiomyocytes, and promoted functional recovery in mice via reducing fibrosis, enhancing the proliferation of cardiomyocytes, and increasing angiogenesis [123].

In addition to taking advantage of the biological functions, EVs have also been used as drug-delivery systems for the treatment of cardiovascular diseases. For example, Yang et al. pretreated MSCs with atorvastatin (ATV), , which is a widely used lipid-lowering drug for coronary heart disease, and isolated exosomes (ATV-MSC-Exo) for acute MI treatment (Figure 4.19) [124]. The results indicated that the ATV-MSC-Exo could remarkably enhance cardiac function and improve blood vessel formation compared to that of normal MSC-Exo. Further analysis revealed that the ATV-MSC-Exo up-regulated the expression level of lncRNA H19. In addition, silencing of lncRNA H19 abrogated the cardioprotective effects of ATV-MSC-Exo, while overexpression of lncRNA H19 achieved similar therapeutic effects as ATV pretreatment. Altogether, this work highlighted that the pretreatment of MSCs with ATV is an efficient method to improve the therapeutic effects of MSC-Exo in MI treatment [124].

Transplanted MSCs are heavily influenced by the local pro-apoptotic microenvironment, resulting in the mass death of most transplanted MSCs [125]. However, the implanted MSCs produced positive therapeutic effects; so it is reasonable to assume that apoptosis may play a role in the therapeutic effects. Jin et al. found

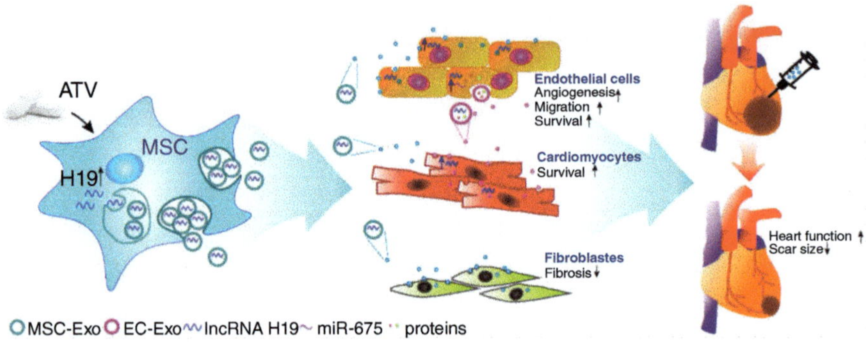

Figure 4.19 Schematic illumination of exosomes derived from Atorvastatin-pretreated MSCs for acute MI treatment. Source: Huang et al. [124]/Oxford University Press.

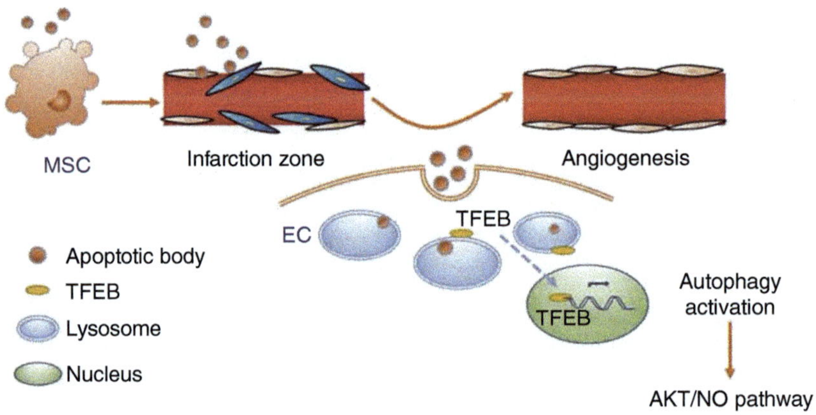

Figure 4.20 Schematic demonstration of the angiogenesis promotion effects of MSC-derived apoptotic bodies via regulation of autophagy in endothelial cells. Source: Liu et al. [126]/Taylor & Francis.

that apoptotic donor MSCs promote angiogenesis via regulating autophagy in the recipient ECs, revealing the role of donor cell apoptosis in the therapeutic effects generated by cell transplantation (Figure 4.20) [126]. The authors first found that implantation of MSCs induced apoptosis while improving cardiac function in rats with MI, suggesting a potential link between MSCs apoptosis and their therapeutic outcomes. Then, they injected apoptotic bodies (Abs) derived from MSCs into MI rats, and found that the AB-treated group exhibited significant left ventricle function improvement. In addition, the infraction zone size and the blood vessel density in the Abs-treated group were significantly higher than that in the PBS-treated group. Furthermore, through analyzing the expression of angiogenesis-associated genes in endothelial cells (ECs) via qPCR, the authors showed that proangiogenic genes (ANGPT1 and KDR) were significantly upregulated, while anti-angiogenic genes (THBS1 and VASH1) were downregulated after treatment. These results indicated that donor MSC-derived Abs could activate the angiogenic capacity of

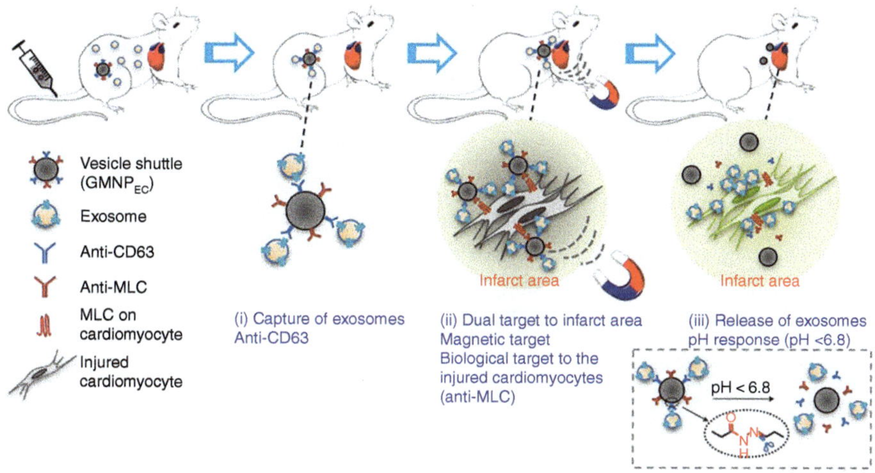

Figure 4.21 Schematic illumination of antibody-functionalized magnetic nanoplatforms for treatment of infarcted heart tissue through the capture and delivery of circulating exosomes. Source: Liu et al. [129]/Springer Nature.

Ecs. They next studied the mechanisms underlying the Abs-induced proangiogenic effects on Ecs. They found that the majority of PKH67-labeled Abs co-localized with lysosomes, and this co-location phenomenon disappeared when Ecs were pretreated with the lysosomal inhibitor chloroquine. Moreover, Abs could enhance lysosomal protease activity. TFEB is a key molecule that regulates lysosomal biogenesis and autophagy [127]. In addition, TFEB has beneficial results on postischemic angiogenesis [128]. After Abs treatment, the relative expression of TFEB in Ecs was upregulated. It suggested that Abs enhanced the biogenesis of lysosomes by activating TFEB. Abs upregulated autophagy and angiogenesis via upregulating TFEB in Ecs. Studies have revealed that apoptosis may be the driving force for cell proliferation during tissue regeneration in many model organisms. In conclusion, MSC-derived Abs can improve the ability of myocardial tissue repair by reducing infarct size and improving cardiac function via angiogenesis in the MI model [126].

In addition, Jin et al. developed a nanoparticle-based platform for on-demand manipulation of exosome distribution (Figure 4.21) [129]. The nanoparticle termed "vesicle shuttle" consisted of a ferro ferric oxide (Fe_3O_4) core for magnetic-field-induced targeting, a silica (SiO_2) shell, and a stimuli-cleavable (PEG) corona conjugated to two antibodies. An anti-CD63 antibody was used to capture the antigens on the exosomes of interest, and an anti-MLC antibody was used to target the injured cardiomyocytes in the MI region. The significant higher signal was detected in the heart after the administration, indicating the capture and transportation ability of the vesicle shuttle. In addition, treatment of the vesicle shuttle significantly reduced the size of MI, a decreased left-ventricle ejection fraction, and promoted angiogenesis in rats compared to that of control groups. This work highlighted a novel concept of vesicle shuttle as a tool to regulate *in vivo* EVs biodistribution for MI treatment [129].

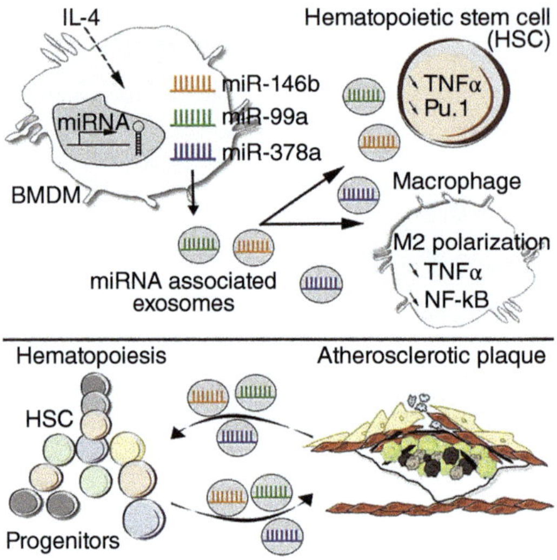

Figure 4.22 Schematic demonstration of exosomes derived from M2-like macrophages induced by IL-4 for treatment of inflammatory disorders. Source: Bouchareychas et al. [130]/with permission of Elsevier.

Also, Raffai et al. showed that exosomes derived from mouse naïve bone marrow-derived macrophages (BMDM-exo) contain anti-inflammatory microRNA-99a/146b/378a (Figure 4.22) [130]. The microRNAs contained in exosomes could be further increased by BMDM polarized with cytokine IL-4 (BMDM-IL-4-exo). These exosomal microRNAs in BMDM-IL-4-exo were demonstrated to participate in the regulation of inflammatory responses via upregulation of NF-κB expression, down-regulation of TNF-α, and promotion of M2 polarization in recipient macrophages. Repeated administrations of BMDM-IL-4-exo into ApoE$^{-/-}$ mice with a western diet reduced hematopoiesis in the bone marrow, myeloid cells in the circulation, and macrophage accumulation in aortic root lesions. The authors also observed that BMDM-IL-4-exo treatment decreased the expression of genes associated with monocyte activation, mobilization, and recruitment to atherosclerotic vessels. This work demonstrated that exosomes produced by IL-4-induced M2-like macrophages could be a novel therapeutic for inflammatory disorders including atherosclerosis [130].

Exosomes derived from cardiosphere-derived cells (CDCs) were demonstrated to show therapeutic potential in improving cardiac hypertrophy and dysfunction in an angiotensin II (Ang II)-induced cardiac hypertrophy mouse model [131]. In addition, heart homing peptide (HHP) has been shown to localize in the heart tissue after intravenous administration via specific binding to cysteine-rich protein 2 (CRIP2) [132]. Accordingly, Wang et al. developed modified exosomes derived from human CDCs with the surface display of HHP (HHP-EXO) for cardiac hypertrophy treatment (Figure 4.23) [133]. The results indicated that HHP-EXO significantly improved cardiac function and reduced fibrosis in a mouse model of myocardial hypertrophy via transverse aorta constriction. In addition, this study showed that miRNA-148a holds a potential contribution in mediating the therapeutic effect of HHP-EXO via down-regulation of GP130, subsequently leading to inhibition of STAT3, ERK1/2, and AKT signaling pathways. In conclusion, this

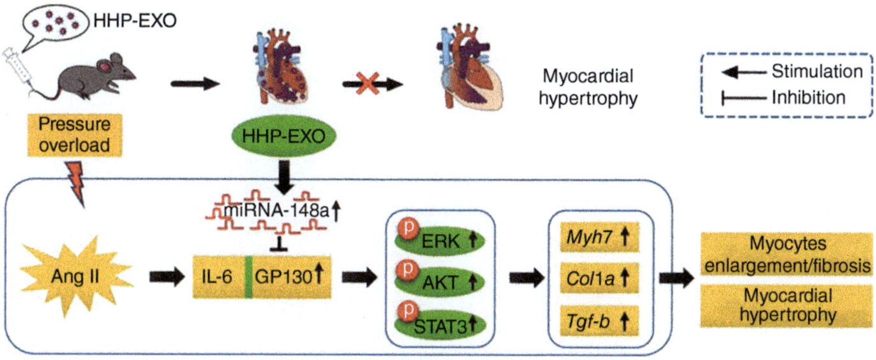

Figure 4.23 Schematic illumination of heart homing peptide-modified exosomes (HHP-EXO) derived from cardiosphere-derived cells for treatment of cardiac hypertrophy. Source: Mao et al. [133]/Springer Nature/CC BY 4.0.

work demonstrated that HHP-modified exosomes could enhance the therapeutic effect of CDCs-exosomes on cardiac hypertrophy [133].

Atherosclerosis (AS), especially coronary atherosclerosis—the major cause of CVD, is a multifocal, smoldering, immunoinflammatory disease of medium-sized and large arteries fueled by lipids [134, 135]. At present, the main clinical treatments of atherosclerosis include lifestyle interventions, lipid-lowering therapies, antithrombotic treatments, and percutaneous intervention treatments [136–138]. Krüppel-like factor 5 (KLF5)—a zinc-finger-containing transcription factor plays a central role in cardiovascular remodeling by mediating smooth muscle cell (SMC) proliferation and migration [139]. Recently, Wen et al. demonstrated that KLF5 exerted its pro-atherosclerotic effect by regulating the expression of multiple miRNAs, most prominently miR-155 (Figure 4.24) [140]. They found that miR-155-rich exosomes were secreted from KLF5-overexpressing vascular smooth muscle cells (VSMCs) and were taken up by recipient ECs [140]. Upon uptake by ECs, miR-155 inhibited the proliferation and migration of ECs, and impaired endothelial barrier function by suppressing the expression of tight junction proteins. Therefore, blockage of the exosome-mediated transfer of miR-155 between these two cells may serve as a therapeutic target for atherosclerosis [140].

In addition, Yuan et al. developed a nano-system for the delivery of inflammation-responsive IL-10 mRNA via exosomes for inflammation control [141]. The exosome-based delivery of engineered IL-10 mRNA alleviated atherosclerosis in the ApoE$^{-/-}$ mice model. This work established a potent platform for tailored inflammation control and a promising strategy for atherosclerosis treatment [141].

4.4 Respiratory Diseases

Respiratory diseases are one of the leading causes of mortality worldwide, causing a heavy burden to society. Though a variety of drugs, such as small molecules, mRNA, proteins, and stem cells, have been developed for treating respiratory diseases,

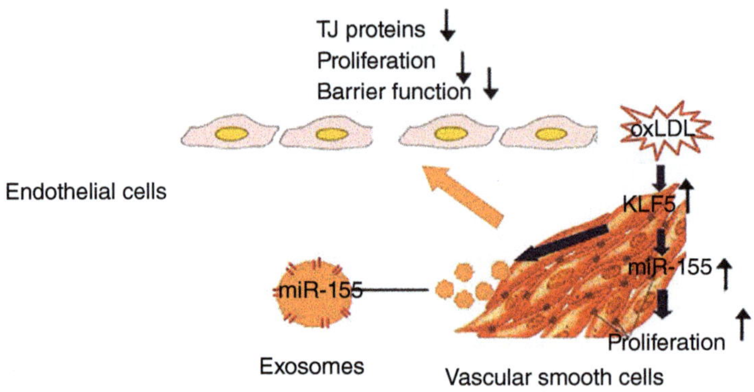

Figure 4.24 Schematic representation of the functions of miR-155-rich exosomes. Source: Zheng et al. [140]/with permission of Elsevier.

The therapeutic efficacy of proposed drugs, including small molecules, nucleic acids, peptides, and cells, is strictly suppressed [142]. EVs represent an alternative technique for the treatment of respiratory diseases.

Chen et al. isolated human umbilical cord mesenchymal stem cells (hucMSCs)-derived exosomes and found that their administration could improve the lung morphometry in lipopolysaccharide (LPS)-induced acute lung injury (ALI) [143]. Further analysis indicated that miR-377-3p overexpressed in the hucMSCs-exosomes play a crucial role in the regulation of autophagy, resulting in protection effects. Compared to exosomes from human fetal lung fibroblast cells (HFL-1), hucMSCs-exosomes indicated superior therapeutic efficacy in the suppression of bronchoalveolar lavage (BALF), inflammatory factors, and induced autophagy. This study suggested that hucMSCs-exosomes provide a potential effective therapeutic approach for the treatment of ALI [143].

In another work, Shi et al. investigated the protection effects of adipose-derived mesenchymal stem cells-derived exosomes (AdMSC-Exos) on ALI [144]. The authors figured out that AdMSC-Exos could efficiently transfer the mitochondrial components into alveolar macrophages, leading to elevated mtDNA level, mitochondrial membrane potential, oxidative phosphorylation activity, ATP generation, and reduced mROS stress in LPS-treated macrophages. The macrophages then shifted to anti-inflammatory phenotype for protective effects. The authors proposed that the AdMSC-Exos could be applied in the treatment of critical lung diseases such as COVID-19 [144].

Recently, Cheng et al. isolated lung spheroid cells-derived secretome (LSC-Sec) and exosomes (LSC-Exo) to attenuate and resolve bleomycin- and silica-induced fibrosis [145]. The results indicated that the LSC-Sec and LSC-Exo treatment could attenuate alveolar epithelial and vascular injury, reduce the fibrotic response, and decrease αSMA^{+} cells. In addition, LSC-Sec and LSC-Exo exhibited superior therapeutic benefits than counterparts derived from MSCs. This work suggested potential candidates for the development of therapeutics for idiopathic pulmonary fibrosis [145].

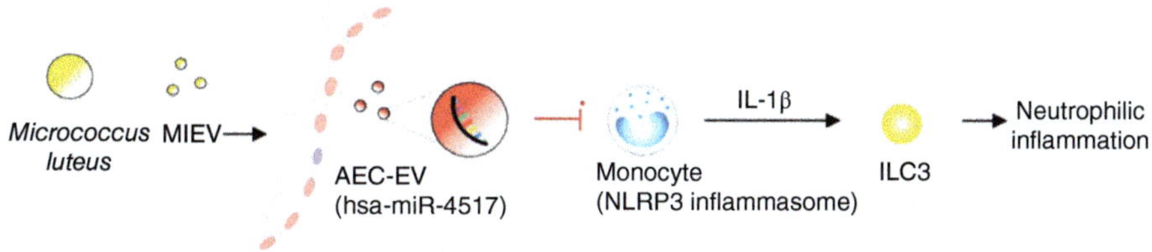

Figure 4.25 Schematic illumination of airway epithelial cells-derived EVs (AEC-EV) for treatment of neutrophilic asthma. Source: Sim et al. [146]/Springer Nature/CC BY 4.0.

Bacterial EVs have demonstrated the potential in regulating pulmonary diseases, nevertheless, their application, and functions in neutrophilic asthma (NA) is yet unclear. To this end, Park et al. investigated the clinical function of *Micrococcus luteus*-derived EVs (MLEVs) in asthma (Figure 4.25) [146]. The data revealed significantly lower MIEV-specific IgG4 in patients with NA compared to that in patients with eosinophilic asthma (EA). In addition, MLEVs could suppress monocyte function in the lung tissues of asthmatic mice via the regulation of miRNAs in airway epithelial cells (AECs). Further analysis indicated that the pivotal role of hsa-miR-4517 in this process by reducing IL-1β production, NLRP3 expression in LPS-stimulated monocytes, and subsequently leading to the suppressed ILC3 activation and neutrophil recruitment. This work suggested that MLEVs could be used as therapeutic agents for the treatment of NA [146].

The world has just suffered from the COVID-19 pandemic that was caused by SARS-CoV-2 coronavirus. Many researchers investigated the applications of EVs in the treatment of COVID-19. For example, Moon et al. demonstrated that MSCs-derived EVs could reverse the SARS-CoV-2-induced pathogenic outcome and inhibit the proinflammatory functions by suppressing the replication of SARS-CoV-2 [147]. EVs-based delivery systems for treating COVID-19 have also been investigated. For example, Gould et al. proposed a novel technique to fabricate exosomes for mRNA delivery [148]. The authors loaded mRNAs that encode SARS-CoV-2 Spike and Nucleocapsid proteins in the exosomes, inducing strong immunity for months. Popowski et al. prepared lung cells-derived exosomes (Lung-Exos) mRNA and protein delivery [149]. More importantly, the developed Lung-Exos could be formulated into lyophilized dry powder with excellent stability. The inhalation of Lung-Exos loaded with SARS-CoV-2 Spike protein-encoding mRNA led to stronger immunity than liposomes. Popowski et al. also systematically compared the biodistribution of Lung-Exos with HEK293T-derived exosomes and synthetic liposomes in the market [150]. The results indicated that the Lung-Exos possess superior properties regarding the mRNA and protein delivery. More exciting, ExoFloTM, which is an exosome agent derived from BM-MSC, has been tested in a nonrandomized open-label cohort study for COVID-19 treatment [151]. The data indicated that ExoFlo™ improved patient recovery by restoring oxygen storage capacity, downregulating cytokine storm, and enhancing immunity.

4.5 Cancers

Cancer is still one leading cause of global human death due to its late diagnosis, poor prognosis, and metastasis [152]. EVs are promising platforms for cancer therapy because of the fascinating functions, such as drug delivery, mediation of cell–cell communication, and induction of immunity [27].

For example, Zhang et al. developed a novel drug delivery system capable of changing and treating hepatocellular carcinoma (HCC) [153]. The previous studies demonstrated the unique biosafety of red blood cell-derived EVs (RBC-EVs) as drug delivery vesicles [154]. The authors here showed that RBC-EVs possess natural liver accumulation. The local environment in the liver induced phagocytization of RBC-EVs by macrophages mediated by C1q. RBC-EVs loaded with doxorubicin (DOX) and sorafenib (SRF) indicated superior anticancer efficacy compared to bare DOX and SRF in a murine model HCC. More importantly, drug-loading RBC-EVs demonstrated remarkable biocompatibility. This work suggested the promising potential of drug-loading RBC-EVs in the treatment of HCC toward clinical applications [153].

In another study, Rivoltini et al. developed exosomes functionalized with TNF-related apoptosis-inducing ligand (TRAIL) for cancer therapy. *In vitro* results demonstrated that $TRAIL^+$ exosomes induced significant cancer cell apoptosis. Intratumoral administration of $TRAIL^+$ exosomes in remarkable suppression of tumor growth in mice. This work suggested that exosome-mediated TRAIL delivery is a potential therapeutic approach for efficient cancer treatment [155].

Bacterial outer membrane vesicles (OMVs) serve as a promising cancer vaccine platform for effective stimulation of anticancer immune responses as OMVs could induce severe toxicity by activating the innate immune system. Park et al. developed a simple isolation technique for the production of artificial OMVs called synthetic bacterial vesicles (SyBV) for cancer immunotherapy (Figure 4.26) [156]. The results indicated that SyBV with excellent biocompatibility induced similar effects in the activation of mouse bone marrow-derived dendritic cells. Administration of SyBV and melanoma EVs induced significant tumor suppression through Th-1 type T cell immunity in mice models. This work suggested a safe and efficient scheme for specific antitumor effects using immunotherapeutic synthetic bacterial OMVs [156].

Lung cancer is a fatal disease with the highest incidence and mortality rate among malignancies [157]. Currently, the available treatment methods are insufficient to improve survival and quality of life in patients with advanced lung cancer. EVs from the tumor or immune cells could mediate immune regulation in the tumor microenvironment. Also, oncolytic viruses exhibit different anticancer mechanisms compared to conventional therapies [158]. The combination of chemotherapeutic drugs and oncolytic viruses exhibited even stronger cytotoxic responses and oncolysis effects [159]. Accordingly, Garofalo et al. developed an EV-based system for targeted delivery of oncolytic adenovirus and paclitaxel [160]. The developed EV formulation significantly inhibited the tumor growth in animal xenograft models of human lung cancer. In addition, the combination therapy of oncolytic adenoviruses and chemotherapeutic drugs indicated stronger antitumor efficacy both *in vitro*

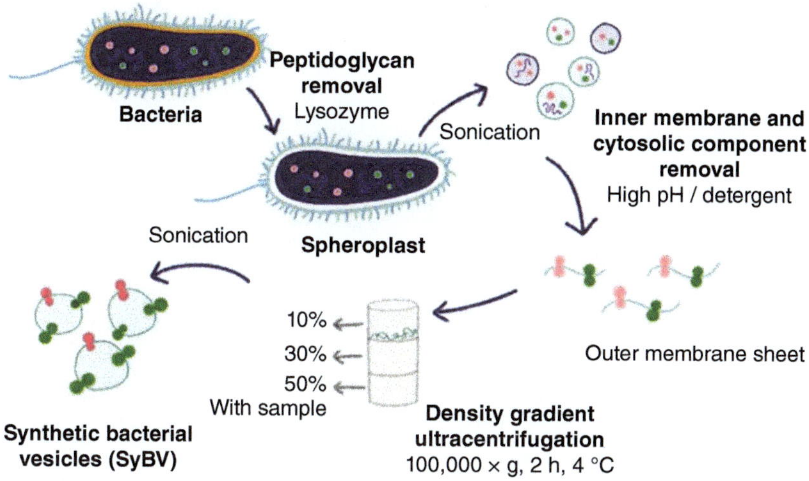

Figure 4.26 Schematic illumination of the preparation of synthetic bacterial outer membrane vesicles. Source: Park et al. [156]/John Wiley & Sons/CC BY 4.0.

and *in vivo*. This study provided a potential strategy for the development of a combination of chemotherapy and viral therapy via the EV-based drug delivery system for lung cancer treatment [160].

In another work, Lin et al. developed another drug delivery system based on bionic artificial exosomes (AEs) that were prepared via the integration of membrane proteins of cancer cells and phospholipid liposomes (Figure 4.27) [161]. The synthetic AEs were used to deliver chemotherapeutic DOX and vorinostat (SAHA), which is an epigenetic inhibitor. The combination of DOX and SAHA demonstrated remarkable synergistic antitumor effects and inhibited the proliferation of non-small cell lung cancer cells and the growth of xenograft tumors without inducing significant side effects. This work highlighted the potential of bioengineered artificial exosomes as novel delivery nanoplatform for the treatment of non-small cell lung cancer [161].

In addition, Kamerkar et al. engineered exosomes derived from fibroblast-like mesenchymal cells (iExosomes) for delivery of siRNAs and shRNAs targeting the mutant form of GTPase KRAS, which is a key driver of pancreatic cancer [162]. Compared to that of lysosomes, the iExosomes indicated enhanced KRAS silencing efficacy in a CD47-dependent manner and is promoted via micropinocytosis process. In addition, the administration of iExosomes significantly inhibited tumor growth in different pancreatic cancer mouse models and improved overall survival. This work demonstrated a novel strategy for specific silencing of oncogenic KRAS for cancer gene therapy [162].

Li et al. developed several strategies for the development of EVs-based cancer vaccines. For example, Fan et al. mimicked the concept of CAR-T cell construction and proposed antibody-engineered exosomes for cancer therapy [163]. The exosomes were harvested from antigen-feeding dendritic cells (tDC-Exo) for providing MHC–antigen complexes and CD86 molecules. The tDC-Exo were further modified with anti-CD3 and anti-EGFR antibodies for promoted binding of T cells to cancer

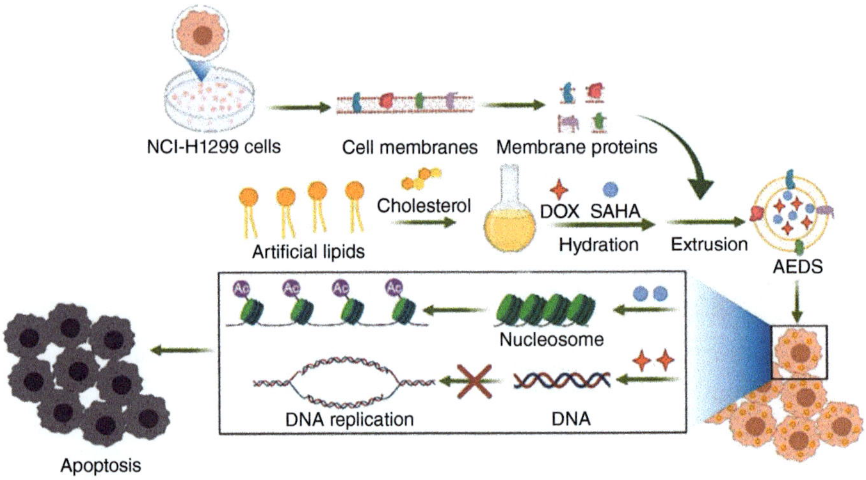

Figure 4.27 Schematic representation of bioinspired artificial exosomes for co-delivery of doxorubicin (DOX) and vorinostat (SAHA) for lung cancer therapy. Source: Lin et al. [161]/with permission of Elsevier.

cells. The developed nanoplatform demonstrated remarkable efficacy in T cell activation for enhanced antitumor effects, providing a novel CAR-T-mimicking strategy for precise cancer immunotherapy [163].

In another study, Zhao et al. developed a novel cancer vaccine based on cancer cell apoptotic bodies (Abs) loaded with CpG oligodeoxyribonucleic acid (CpG ODN) (Figure 4.28) [164]. The proposed Abs-CpG could target and polarize macrophages to the M1 phonotype, and significantly improve the immunosuppressive tumor microenvironment. In addition, the *in vitro* and *in vivo* results indicated that the Abs-CpG could induce strong immune responses and remarkably suppress tumor growth, recurrence, and metastasis. This study proposed a novel strategy for the development of an Abs-based cancer vaccine platform for efficient cancer immunotherapy [164].

Nie et al. also reported the construction of responsive exosome nano-bioconjugates for cancer immunotherapy [165]. In general, exosomes derived from M1 macrophages were modified with azidine group and further conjugated with anti-CD47 (aCD47) and anti-SIRPα (aSIRPα) antibodies with dibenzocyclooctyne modification. The developed exosomal nanoparticle demonstrated aCD47-dependent targeting capability, efficient blocking of the "don't eat me" signaling pathway via aCD47 and aSIRPα, and reprogramming of M2 to M1 phonotype for potent anticancer therapeutic effects [165].

Hong et al. developed an enzymatic exosome loaded with native glycosylphosphatidylinositol-anchored form of hyaluronidase (Exo-PH20) for anticancer therapy [166]. The Exo-PH20 demonstrated enhanced tumor and drug diffusion into tumor foci through hyaluronan degradation. Furthermore, the codelivery of PH20 hyaluronidase and DOX by developed exosomes significantly inhibited tumor growth and enhanced T cell infiltration in the tumor. This study provides

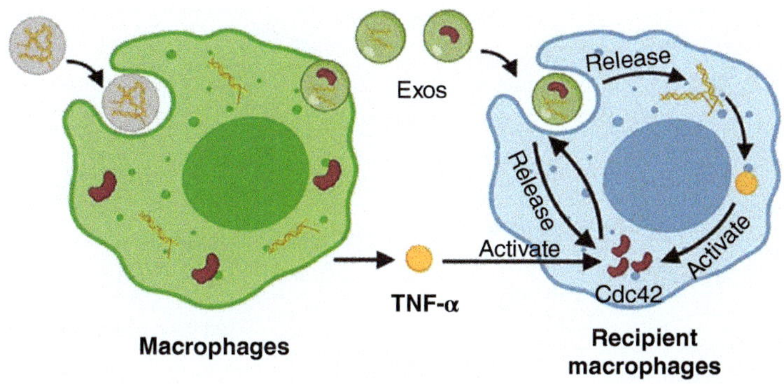

Figure 4.28 Schematic demonstration of exosomes derived from M1 macrophage for CpG ODN delivery for cancer immunotherapy. Source: Zhao et al. [164]/with permission of Elsevier.

a promising exosome-based nanoplatform with high enzyme activity and drug loading efficiency [166].

Moreover, Zhao, et al. developed a novel type of exosome-based biomimetic nanoparticles for triple-negative breast cancer therapy [167]. The nanoplatform was established via conjugation of cationic bovine serum albumin (CBSA) and S100A4 siRNA (siS100A4), and further coating by exosome membrane (CBSA/siS100A4@Exosome). The developed CBSA/siS100A4@Exosome demonstrated superior targeting to the lung and exhibited remarkable gene silencing efficacy, leading to the inhibition of the growth of malignant tumor. This study represented a promising exosome-based strategy for gene therapy of breast cancer metastasis [167].

Recently, Jiao, et al. proposed a novel EVs-based approach for efficient delivery of triphenylphosphonium (TPP)-modified P53 proteins (TPP/P53@EVs) for breast cancer treatment (Figure 4.29) [168]. EVs derived from breast cancer cells were selected as vesicles due to their homing capability to patent cells. Indeed, the authors observed the uptake and accumulation of the TPP/P53@EVs into breast cancer cells and tumors after systematic administration. In addition, the theranostic nanoplatform demonstrated excellent inhibition of tumor growth via the interaction with Bcl-2 family members and induction of cancer cell apoptosis. This work provided a "drug-free" strategy for breast cancer therapy toward future clinical applications [168].

Glioblastoma (GBM) is one of the most aggressive tumors with a poor prognosis in the central nervous system tumors [169]. To enhance the blood–brain barrier (BBB) penetration of drugs, Li et al. constructed a novel drug delivery system via the conjugation of magnetic nanoparticles (MNPs) and engineered exosomes with siRNA encapsulation [170]. The MNPs consisted of Fe_3O_4 core and mesoporous silica shell with further modification with anti-CD63 antibody for specific binding of EVs. In addition, the mesoporous silica shell was conjugated with brequinar (BQR), which is an inhibitor of dihydroorotate dehydrogenase (DHODH) in clinical use. The hMSCs-derived EVs were decorated with ANG peptide (ANG-EXO) via

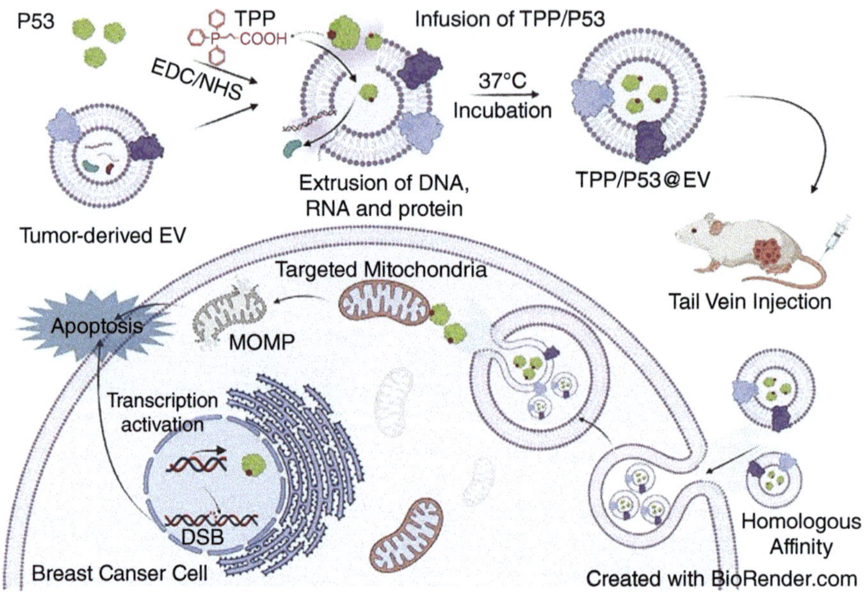

Figure 4.29 Schematic illumination of tumor cell-derived EVs for the delivery of P53 proteins for breast cancer therapy. Source: Jiao et al. [168]/with permission of Elsevier.

incorporation of ANG and Lamp2b fusion protein, and further loaded with siRNA of GPX4 (siGPX4). The resulting nanoplatforms were demonstrated to accumulate in the brain under a magnetic field and penetrated the BBB via ANG recognition of LRP-1 receptors. *In vitro* and *in vivo* experiments revealed the synergistic ferroptosis therapy of GBM due to the combination activities of DHODH disintegration, suppressed GPX4 ferroptosis defense pathway, and Fe^{2+} release. This study provided novel EVs-based nanoplatforms for enhanced synergistic ferroptosis therapy of GBM [170].

Additionally, sonodynamic therapy (SDT) has also been accessed as a potential treatment modality for GBM due to its high tissue penetration and negligible radiation damage to normal tissues. Nevertheless, the BBB and the hypoxic microenvironment limit the efficiency of SDT [171]. Accordingly, Wu et al. proposed a novel biodegradable nanomedicine with properties of efficient BBB penetration and excellent tumor targeting [172]. The nanoplatform consisted of silica nanoparticles with catalase (CAT) encapsulation and sonosensitizer indocyanine green (ICG) loading. The nanoplatform was further coated with macrophage exosomes that were modified with AS1411 aptamers. The cellular glutathione could induce the degradation of the nanoplatform, and the released CAT would catalyze the O_2 production for the relief of tumor hypoxic microenvironment, enhancing the SDT efficiency against GBM. This work suggested a promising nanoplatform for GBM SDT therapy with the potential in clinical translation [172].

In another study, McDonald et al. developed novel engineered exosomes for delivery of miRNAs for targeted GBM treatment. Compared with natural exosomes, the engineered exosomes demonstrated superior efficiency in the loading and delivery of anti-glioma nucleic acids and therapeutic efficacy in the treatment of GBM [173]. In addition, Huang et al. designed engineered exosomes with c(RGDyK) modification and lncRNA MEG3 encapsulation (cRGD-Exo-MEG3). The developed cRGD-Exo-MEG3 demonstrated excellent anti-osteosarcoma (OS) efficacy both *in vitro* and *in vivo*, providing an efficient EVs-based drug delivery strategy for OS therapy [174]. Moreover, Han et al. modified 293T-derived exosomes with cholesterol-covalent E3 aptamer and loaded SIRT6-siRNA into the engineered exosomes for prostate cancer therapy [175].

4.6 Conclusion and Perspectives

Increased understanding of the physiological and pathological properties of extracellular vesicles (EVs) brings novel possibilities in nanomedicine, drug delivery, and disease therapy [5, 176]. Indeed, massive studies already demonstrated the promising potential of EVs in these fields due to their excellent biosafety, targeting ability, and easy functionality [177]. Nevertheless, there are still some disadvantages of EVs hampering the clinical translation and industrial applications of EVs-based reagents.

Large-scale preparation remains the major challenge for EV translation. The current isolation and purification methods can't satisfy the demands of abundant EVs for different applications [178]. It was proposed that classical manufacturing skills, such as protein or cell production, might inspire the large-scale production of EVs. Therefore, large bioreactors, including cartridges, wave bags, and flasks, have been suggested for cell culture [5]. In addition, researchers have proposed many schemes for the generation of artificial EV-like nanovesicles. For example, Jang and others came up with the extrusion-based technique for the massive production of nanovesicles with similar properties as natural EVs but significantly elevated yield (over 20-fold) [53, 179]. In addition, other methods, such as filtration-based techniques et al., offer various choices for the production of EV-like nanovesicles [180]. The inherent heterogeneity of EVs also limits the applications of EVs. Some strategies, such as chromatography et al., were applied for the purification of EVs [177]. Despite the success, it is still necessary to develop novel techniques for the large-scale generation and purification of functional EVs. Insufficient therapeutic efficacy is another challenge for EV translation. Though EVs contain various biomolecules and thus regulate many biological processes, the drug loading efficiency is relatively low. In addition, the targeted drug delivery of EVs is not satisfactory [6, 12]. Different strategies, including genetic bioengineering, chemical modification, and metabolic labeling, are proposed to improve the targeted distribution of EVs [10, 89]. But we need to re-evaluate the negative influences of the modifications on engineered EVs. In addition, the therapeutic efficacy and biosafety should be carefully investigated before the clinical translation. Last but not least, the bioavailability of EVs restricted the therapeutic outcome of EVs. To achieve the maximal therapeutic effects, different biomaterials are combined with EVs to improve their retention in targeted tissues [11]. Accordingly, the detailed properties of material-based EV systems should be well studied.

EVs have been considered promising candidates in the development of next-generation nanomedicine. The current EVs-based nanoplatforms are not fully prepared for clinical translation due to the above-discussed issues. Nevertheless, future advances in biology, chemistry, and nanomedicine would assuredly pave new paths for the development and applications of EVs. We hold strong faith in the clinical translation of EVs in the future.

References

1 Yuana, Y., Sturk, A., and Nieuwland, R. (2013). *Blood Rev.* 27: 31–39.
2 Yáñez-Mó, M., Siljander, P.R.-M., Andreu, Z. et al. (2015). *J. Extracell. Vesicles* 4: 27066.
3 Malloci, M., Perdomo, L., Veerasamy, M. et al. (2019). *Antioxid. Redox Signaling* 30: 813–856.
4 Mentkowski, K.I., Snitzer, J.D., Rusnak, S., and Lang, J.K. (2018). *AAPS J.* 20: 1–17.
5 Herrmann, I.K., Wood, M.J.A., and Fuhrmann, G. (2021). *Nat. Nanotechnol.* 16: 748–759.
6 Vader, P., Mol, E.A., Pasterkamp, G., and Schiffelers, R.M. (2016). *Adv. Drug Delivery Rev.* 106: 148–156.
7 Gowen, A., Shahjin, F., Chand, S. et al. (2020). *Front. Cell Dev. Biol.* 8: 149.
8 Aghajani Nargesi, A., Lerman, L.O., and Eirin, A. (2017). *Stem Cell Res. Ther.* 8: 1–12.
9 Armstrong, J.P., Holme, M.N., and Stevens, M.M. (2017). *Acs Nano* 11: 69–83.
10 Richter, M., Vader, P., and Fuhrmann, G. (2021). *Adv. Drug Delivery Rev.* 173: 416–426.
11 Murali, V.P. and Holmes, C.A. (2021). *Acta Biomater.* 124: 88–107.
12 Elkhoury, K., Koçak, P., Kang, A. et al. (2020). *Pharmaceutics* 12: 849.
13 Zhang, X., Zhang, H., Gu, J. et al. (2021). *Adv. Mater.* 33: 2005709.
14 Sahoo, S., Adamiak, M., Mathiyalagan, P. et al. (2021). *Circulation* 143: 1426–1449.
15 de Abreu, R.C., Fernandes, H., da Costa Martins, P.A. et al. (2020). *Nat. Rev. Cardiol.* 17: 685–697.
16 Reiner, A.T., Witwer, K.W., Van Balkom, B.W. et al. (2017). *Stem Cells Trans. Med.* 6: 1730–1739.
17 Möller, A. and Lobb, R.J. (2020). *Nat. Rev. Cancer* 20: 697–709.
18 Boulanger, C.M., Loyer, X., Rautou, P.-E., and Amabile, N. (2017). *Nat. Rev. Cardiol.* 14: 259–272.
19 Elsharkasy, O.M., Nordin, J.Z., Hagey, D.W. et al. (2020). *Adv. Drug Delivery Rev.* 159: 332–343.
20 Wu, P., Zhang, B., Ocansey, D.K.W. et al. (2021). *Biomaterials* 269: 120467.
21 Yamamoto, T., Kosaka, N., and Ochiya, T. (2019). *Sci. Technol. Adv. Mater.* 20: 746–757.

22 Borrelli, D.A., Yankson, K., Shukla, N. et al. (2018). *J. Controlled Release* 273: 86–98.
23 Witwer, K.W., Van Balkom, B.W., Bruno, S. et al. (2019). 1609206. *J. Extracell. Vesicles* 8.
24 Piffoux, M., Volatron, J., Cherukula, K. et al. (2021). *Adv. Drug Delivery Rev.* 178: 113972.
25 Walker, S., Busatto, S., Pham, A. et al. (2019). *Theranostics* 9: 8001.
26 Kalluri, R. and LeBleu, V.S. (2020). *Science* 367: eaau6977.
27 Wu, Q., Fu, S., Xiao, H. et al. (2023). *Adv. Sci.* 10: 2204814.
28 Xie, J., Li, Q., Haesebrouck, F. et al. (2022). *Trends Biotechnol.* 40: 1173–1194.
29 Liu, H., Zhang, Q., Wang, S. et al. (2022). *Bioact. Mater.* 14: 169–181.
30 Anand, S., Samuel, M., and Mathivanan, S. (2021). *New Frontiers: Extracellular Vesicles*, 89–97. Springer.
31 Katsuda, T., Oki, K., and Ochiya, T. (2015). *Stem Cell Renewal and Cell-Cell Communication: Methods and Protocols*, 171–181. Springer.
32 Murugan, B. and Sagadevan, S. (2022). *Extracellular Vesicles: Role in Diseases, Pathogenesis and Therapy*, 309. IntechOpen.
33 Federico, M. and Ridolfi, B. (2022). *Extracellular Vesicles in Diagnosis and Therapy*, 219–230. Springer.
34 Thompson, A.G., Gray, E., Heman-Ackah, S.M. et al. (2016). *Nat. Rev. Neurol.* 12: 346–357.
35 Maas, S.L., Breakefield, X.O., and Weaver, A.M. (2017). *Trends Cell Biol.* 27: 172–188.
36 Huang, K., Hu, S., and Cheng, K. (2019). *Adv. Healthc. Mater.* 8: 1801011.
37 Hu, G.-w., Li, Q., Niu, X. et al. (2015). *Stem Cell Res. Ther.* 6: 1–15.
38 Wiklander, O.P., Brennan, M.Á., Lötvall, J. et al. (2019). *Sci. Transl. Med.* 11: eaav8521.
39 Rao, S.-S., Hu, Y., Xie, P.-L. et al. (2018). *Bone Res.* 6: 9.
40 Rachner, T.D., Khosla, S., and Hofbauer, L.C. (2011). *Lancet* 377: 1276–1287.
41 Qi, X., Zhang, J., Yuan, H. et al. (2016). *Int. J. Biol. Sci.* 12: 836.
42 Roubelakis, M.G., Pappa, K.I., Bitsika, V. et al. (2007). *Stem Cells Dev.* 16: 931–952.
43 Chen, C.-Y., Rao, S.-S., Tan, Y.-J. et al. (2019). *Bone Res.* 7: 18.
44 Huang, H., Ma, S., Xing, X. et al. (2023). *Acta Biomater.* 157: 609–624.
45 Ma, L., Ke, W., Liao, Z. et al. (2022). *Bioact. Mater.* 17: 425–438.
46 Tsiapalis, D. and O'Driscoll, L. (2020). *Cells* 9: 991.
47 Jiang, X., You, L., Zhang, Z. et al. (2021). *Front. Cell Dev. Biol.* 9: 693534.
48 Adriano, B., Cotto, N.M., Chauhan, N. et al. (2021). *Bioact. Mater.* 6: 2479–2490.
49 Kim, H., Kim, D.E., Han, G. et al. (2022). *Adv. Healthc. Mater.* 11: 2102027.
50 Little, M.H. and Kairath, P. (2016). *Kidney Int.* 90: 289–299.
51 Ko, K.-W., Park, S.-Y., Lee, E.H. et al. (2021). *Acs Nano* 15: 7575–7585.
52 Gallet, R., Dawkins, J., Valle, J. et al. (2017). *Eur. Heart J.* 38: 201–211.
53 Wang, X., Hu, S., Li, J. et al. (2021). *ACS Appl. Mater. Interfaces* 13: 55767–55779.
54 Huang, X., Sun, J., Chen, G. et al. (2019). *Front. Pharmacol.* 10: 421.

55 Wang, Y., Cao, Z., Wei, Q. et al. (2022). *Acta Biomater.* 147: 342–355.
56 Lee, B., Cripps, R.A., Fitzharris, M., and Wing, P. (2014). *Spinal Cord* 52: 110–116.
57 David, G., Mohammadi, S., Martin, A.R. et al. (2019). *Nat. Rev. Neurol.* 15: 718–731.
58 Kim, D.-k., Nishida, H., An, S.Y. et al. (2016). *Proc. Natl. Acad. Sci.* 113: 170–175.
59 Swanson, W.B., Gong, T., Zhang, Z. et al. (2020). *J. Controlled Release* 324: 679–694.
60 Sun, H.H., Jin, T., Yu, Q., and Chen, F.M. (2011). *J. Tissue Eng. Regener. Med.* 5: e1–e16.
61 Zhang, S., Yang, Y., Jia, S. et al. (2020). *Theranostics* 10: 5914.
62 Wang, D., Lyu, Y., Yang, Y. et al. (2022). *Acta Biomater.* 140: 610–624.
63 Kwon, T., Lamster, I.B., and Levin, L. (2021). *Int. Dent. J.* 71: 462–476.
64 Zhang, Y.-F., Shi, J.-B., and Li, C. (2019). *Cytotherapy* 21: 1122–1136.
65 Zarubova, J., Hasani-Sadrabadi, M.M., Dashtimoghadam, E. et al. (2022). *Adv. Healthc. Mater.* 11: 2102593.
66 Cekici, A., Kantarci, A., Hasturk, H., and Van Dyke, T.E. (2014). *Periodontology* 64: 57–80.
67 Shang, F., Yu, Y., Liu, S. et al. (2021). *Bioact. Mater.* 6: 666–683.
68 Lelek, J. and Zuba-Surma, E.K. (2020). *Int. J. Mol. Sci.* 21: 799.
69 Zhang, Q.-Y., Tan, J., Nie, R. et al. (2023). *Composites Part B* 254: 110550.
70 Zhang, L., Ma, S., Wei, P. et al. (2021). *Adv. Healthc. Mater.* 10: 2101298.
71 Pegtel, D.M. and Gould, S.J. (2019). *Annu. Rev. Biochem.* 88: 487–514.
72 Garn, H., Bahn, S., Baune, B.T. et al. (2016). *J. Allergy Clin. Immunol.* 138: 47–56.
73 Mo, R., Jiang, T., Di, J. et al. (2014). *Chem. Soc. Rev.* 43: 3595–3629.
74 Buzási, K., Sápi, Z., and Jermendy, G. (2011). *Diabetes Res. Clin. Pract.* 94: e34–e36.
75 Peyrot, M., Barnett, A., Meneghini, L., and Schumm-Draeger, P.M. (2012). *Diabetic Med.* 29: 682–689.
76 Somiya, M., Yoshioka, Y., and Ochiya, T. (2018). 1440132. *J. Extracell. Vesicles* 7.
77 Wu, L., Wang, L., Liu, X. et al. (2022). *Acta Pharma. Sin. B* 12: 2029–2042.
78 Moreira, A., Kahlenberg, S., and Hornsby, P. (2017). *J. Mol. Endocrinol.* 59: R109.
79 Hameed, I., Masoodi, S.R., Mir, S.A. et al. (2015). *World J. Diabetes* 6: 598.
80 Salem, Z.A., Kamel, A.H.M., and AbuBakr, N. (2021). *J. Mol. Histol.* 52: 467–477.
81 Prentki, M., Matschinsky, F.M., and Madiraju, S.M. (2013). *Cell Metab.* 18: 162–185.
82 Diedisheim, M., Oshima, M., Albagli, O. et al. (2018). *Mol. Metabol.* 10: 74–86.
83 Chen, H., Zhou, W., Ruan, Y. et al. (2018). *Mol. Med.* 24: 1–12.
84 He, Q., Song, J., Cui, C. et al. (2021). *Stem Cell Res. Ther.* 12: 1–16.
85 Dede, A.D., Tournis, S., Dontas, I., and Trovas, G. (2014). *Metabolism* 63: 1480–1490.

86 Bouler, J.-M., Pilet, P., Gauthier, O., and Verron, E. (2017). *Acta Biomater.* 53: 1–12.

87 Jang, J.-W., Yun, J.-H., Lee, K.-I. et al. (2012). *Oral Surg. Oral Med.Oral Pathol. Oral Radiol.* 113: 480–487.

88 Tao, S.-C., Li, X.-R., Wei, W.-J. et al. (2022). *Biomaterials* 283: 121465.

89 Wang, B.Z., Luo, L.J., and Vunjak-Novakovic, G. (2022). *Adv. Healthc. Mater.* 11: 2101557.

90 Zhang, L., Zhu, X.-Y., Zhao, Y. et al. (2020). *Basic Res. Cardiol.* 115: 1–14.

91 Balint, G. and Szebenyi, B. (1996). *Drugs* 52: 1–13.

92 Pereira, D., Peleteiro, B., Araujo, J. et al. (2011). *Osteoarthr. Cartil.* 19: 1270–1285.

93 Yu, S.P. and Hunter, D.J. (2016). *Expert Opin. Pharmacother.* 17: 2057–2071.

94 Thind, A. and Wilson, C. (2016). *J. Extracell. Vesicles* 5: 31292.

95 Rong, Y., Zhang, J., Jiang, D. et al. (2021). *Acta Biomater.* 122: 325–342.

96 Blüher, M. (2019). *Nat. Rev. Endocrinol.* 15: 288–298.

97 Singh, G.M., Danaei, G., Farzadfar, F. et al. (2013). *PLoS One* 8: e65174.

98 Bouter, K.E., van Raalte, D.H., Groen, A.K., and Nieuwdorp, M. (2017). *Gastroenterology* 152: 1671–1678.

99 Zhang, H., Sparks, J.B., Karyala, S.V. et al. (2015). *ISME J.* 9: 770–781.

100 Ashrafian, F., Shahriary, A., Behrouzi, A. et al. (2019). *Front. Microbiol.* 10: 2155.

101 Bagno, L., Hatzistergos, K.E., Balkan, W., and Hare, J.M. (2018). *Mol. Ther.* 26: 1610–1623.

102 Petrie, J.R., Guzik, T.J., and Touyz, R.M. (2018). *Can. J. Cardiol.* 34: 575–584.

103 Fuchs, F.D. and Whelton, P.K. (2020). *Hypertension* 75: 285–292.

104 Dahlöf, B. (2010). *Am. J. Cardiol.* 105: 3A–9A.

105 Jenča, D., Melenovský, V., Stehlik, J. et al. (2021). *Cric.: Heart Failure* 8: 222–237.

106 Reed, G.W., Rossi, J.E., and Cannon, C.P. (2017). *Lancet* 389: 197–210.

107 Zhou, S.-s., Jin, J.-p., Wang, J.-q. et al. (2018). *Acta Pharmacol. Sin.* 39: 1073–1084.

108 Romaine, S.P., Tomaszewski, M., Condorelli, G., and Samani, N.J. (2015). *Heart* 101: 921–928.

109 Boon, R.A. and Dimmeler, S. (2015). *Nat. Rev. Cardiol.* 12: 135–142.

110 Wei, Z., Chen, Z., Zhao, Y. et al. (2021). *Biomaterials* 275: 121000.

111 Fernandes, S., Chong, J.J., Paige, S.L. et al. (2015). *Stem Cell Rep.* 5: 753–762.

112 Wu, Q., Wang, J., Tan, W.L.W. et al. (2020). *Cell Death Dis.* 11: 354.

113 Wang, Y., Li, C., Zhao, R. et al. (2021). *Theranostics* 11: 6315.

114 Negahdaripour, M., Owji, H., Eskandari, S. et al. (2021). *Expert Opin. Biol. Ther.* 21: 371–394.

115 Li, Q., Xu, Y., Lv, K. et al. (2021). *Sci. Transl. Med.* 13: eabb0202.

116 Zhao, J., Li, X., Hu, J. et al. (2019). *Cardiovasc. Res.* 115: 1205–1216.

117 Kudlik, G., Hegyi, B., Czibula, Á. et al. (2016). *Exp. Cell. Res.* 348: 36–45.

118 Hayakawa, K., Esposito, E., Wang, X. et al. (2016). *Nature* 535: 551–555.

119 Ikeda, G., Santoso, M.R., Tada, Y. et al. (2021). *J. Am. Coll. Cardiol.* 77: 1073–1088.

120 Hu, S., Li, Z., Shen, D. et al. (2021). *Nat. Biomed. Eng.* 5: 1174–1188.

121 Qiao, L., Hu, S., Liu, S. et al. (2019). *J. Clin. Invest.* 129: 2237–2250.

122 Zhu, D., Liu, S., Huang, K. et al. (2022). *Circ. Res.* 131: e135–e150.

123 Vandergriff, A., Huang, K., Shen, D. et al. (2018). *Theranostics* 8: 1869.

124 Huang, P., Wang, L., Li, Q. et al. (2020). *Cardiovasc. Res.* 116: 353–367.

125 Bian, S., Zhang, L., Duan, L. et al. (2014). *J. Mol. Med.* 92: 387–397.

126 Liu, H., Liu, S., Qiu, X. et al. (2020). *Autophagy* 16: 2140–2155.

127 Settembre, C., Di Malta, C., Polito, V.A. et al. (2011). *Science* 332: 1429–1433.

128 Fan, Y., Lu, H., Liang, W. et al. (2018). *Circ. Res.* 122: 945–957.

129 Liu, S., Chen, X., Bao, L. et al. (2020). *Nat. Biomed. Eng.* 4: 1063–1075.

130 Bouchareychas, L., Duong, P., Covarrubias, S. et al. (2020). *Cell Rep.* 32: 107881.

131 Cambier, L., Giani, J.F., Liu, W. et al. (2018). *Hypertension* 72: 370–380.

132 Zhang, L., Hoffman, J.A., and Ruoslahti, E. (2005). *Circulation* 112: 1601–1611.

133 Mao, L., Li, Y.-D., Chen, R.-L. et al. (2022). *J. Nanobiotechnol.* 20: 1–14.

134 Björkegren, J.L. and Lusis, A.J. (2022). *Cell* .

135 Stone, G.W., Maehara, A., Lansky, A.J. et al. (2011). *New Engl. J. Med.* 364: 226–235.

136 Charo, I.F. and Taub, R. (2011). *Nat. Rev. Drug Discovery* 10: 365–376.

137 Chen, W., Schilperoort, M., Cao, Y. et al. (2022). *Nat. Rev. Cardiol.* 19: 228–249.

138 Malekmohammad, K., Sewell, R.D., and Rafieian-Kopaei, M. (2019). *Biomolecules* 9: 301.

139 Nagai, R., Suzuki, T., Aizawa, K. et al. (2005). *J. Thromb. Haemost.* 3: 1569–1576.

140 Zheng, B., Yin, W.-n., Suzuki, T. et al. (2017). *Mol. Ther.* 25: 1279–1294.

141 Bu, T., Li, Z., Hou, Y. et al. (2021). *Theranostics* 11: 9988.

142 Popowski, K.D., Dinh, P.U.C., George, A. et al. (2021). *View* 2: 20200186.

143 Wei, X., Yi, X., Lv, H. et al. (2020). *Cell Death Dis.* 11: 657.

144 Xia, L., Zhang, C., Lv, N. et al. (2022). *Theranostics* 12: 2928.

145 Dinh, P.-U.C., Paudel, D., Brochu, H. et al. (2020). *Nat. Commun.* 11: 1064.

146 Sim, S., Lee, D.-H., Kim, K.-s. et al. (2023). *Exp. Mol. Med.* 1–9.

147 Park, J.H., Choi, Y., Lim, C.-W. et al. (2021). *Cells* 10: 2393.

148 Tsai, S.J., Atai, N.A., Cacciottolo, M. et al. (2021). *J. Biol. Chem.* 297.

149 Popowski, K.D., Moatti, A., Scull, G. et al. (2022). *Matter* 5: 2960–2974.

150 Popowski, K.D., de Juan Abad, B.L., George, A. et al. (2022). *J. Extracell. Vesicle* 1: 100002.

151 Sengupta, V., Sengupta, S., Lazo, A. et al. (2020). *Stem Cells Dev.* 29: 747–754.

152 Valkenburg, K.C., De Groot, A.E., and Pienta, K.J. (2018). *Nat. Rev. Clin. Oncol.* 15: 366–381.

153 Zhang, G., Huang, X., Xiu, H. et al. (2020). *J. Extracell. Vesicles* 10: e12030.

154 Biagiotti, S., Abbas, F., Montanari, M. et al. (2023). *Pharmaceutics* 15: 365.

155 Rivoltini, L., Chiodoni, C., Squarcina, P. et al. (2016). *Clin. Cancer Res.* 22: 3499–3512.

156 Park, K.S., Svennerholm, K., Crescitelli, R. et al. (2021). *J. Extracell. Vesicles* 10: e12120.

157 Schabath, M.B. and Cote, M.L. (2019). *Cancer Epidemiol. Biomarkers Prev.* 28: 1563–1579.

158 Bommareddy, P.K., Shettigar, M., and Kaufman, H.L. (2018). *Nat. Rev. Immunol.* 18: 498–513.

159 Ottolino-Perry, K., Diallo, J.-S., Lichty, B.D. et al. (2010). *Mol. Ther.* 18: 251–263.

160 Garofalo, M., Saari, H., Somersalo, P. et al. (2018). *J. Controlled Release* 283: 223–234.

161 Lin, Y., Li, S., Xiao, Z. et al. (2021). *Pharmacol. Res.* 171: 105787.

162 Kamerkar, S., LeBleu, V.S., Sugimoto, H. et al. (2017). *Nature* 546: 498–503.

163 Fan, M., Liu, H., Yan, H. et al. (2022). *Biomaterials* 282: 121424.

164 Zhao, G., Liu, H., Wang, Z. et al. (2022). *Acta Biomater.* 153: 529–539.

165 Nie, W., Wu, G., Zhang, J. et al. (2020). *Angew. Chem. Int. Ed.* 59: 2018–2022.

166 Hong, Y., Nam, G.H., Koh, E. et al. (2018). *Adv. Funct. Mater.* 28: 1703074.

167 Zhao, L., Gu, C., Gan, Y. et al. (2020). *J. Controlled Release* 318: 1–15.

168 Jiao, Y., Tang, Y., Li, Y. et al. (2022). *J. Controlled Release* 349: 606–616.

169 Janjua, T.I., Rewatkar, P., Ahmed-Cox, A. et al. (2021). *Adv. Drug Delivery Rev.* 171: 108–138.

170 Li, B., Chen, X., Qiu, W. et al. (2022). *Adv. Sci.* 9: 2105451.

171 Bilmin, K., Kujawska, T., and Grieb, P. (2019). *Cells* 8: 1428.

172 Wu, T., Liu, Y., Cao, Y., and Liu, Z. (2022). *Adv. Mater.* 34: 2110364.

173 McDonald, M.F., Hossain, A., Momin, E.N. et al. (2020). *Neurosurgery* 67.

174 Huang, X., Wu, W., Jing, D. et al. (2022). *J. Controlled Release* 343: 107–117.

175 Han, Q., Xie, Q.R., Li, F. et al. (2021). *Cancer Res.* 81: 1152–1152.

176 Fu, P., Zhang, J., Li, H. et al. (2021). *Adv. Drug Delivery Rev.* 179: 113910.

177 Li, Y.-J., Wu, J.-Y., Liu, J. et al. (2021). *J. Nanobiotechnol.* 19: 1–20.

178 Akbar, A., Malekian, F., Baghban, N. et al. (2022). *Cells* 11: 186.

179 Jang, S.C., Kim, O.Y., Yoon, C.M. et al. (2013). *Acs Nano* 7: 7698–7710.

180 Leggio, L., Arrabito, G., Ferrara, V. et al. (2020). *Adv. Healthc. Mater.* 9: 2000731.

5

Applications of Engineered Extracellular Vesicles

Lanya Li[1,2], *Yingxian Xiao*[1,2], *Shushan Mo*[3], *and Zhenhua Li*[1,2]

[1] *The Tenth Affiliated Hospital of Southern Medical University, 78 Wanjiang Avenue, Dongguan, Guangdong 523059, China*
[2] *Guangdong Provincial Key Laboratory of Shock and Microcirculation, Shatai South Avenue, Guangzhou, Guangdong 510080, China*
[3] *Hebei University, College of Pharmaceutical Science, Key Laboratory of Pharmaceutical Quality Control of Hebei Province, Wusi East Avenue, Baoding, Hebei, 071002, China*

Even though native Extracellular vesicles (EVs) showed great potential as drug delivery vehicles, diagnostic tools, and even as therapeutic agents themselves, the poor targeting efficiency, low content of functional cargos, heterogeneity, and compositional complexity limit their clinical application [1, 2]. With the use of bioengineering technology, native EVs can be modified to improve their loading efficiency, targeting ability, and therapeutic effect. Engineered EVs inherit the unique advantages of parental cells, such as long cycle times and low immune clearance rates. The engineering strategy for EVs lies in two different ways: loading cargo and modifying the EVs membrane surface (Figure 5.1) [4].

5.1 Engineering EVs for Cargo Loading

Serve as natural nanovehicles, EVs can be used as drug delivery carriers to deliver nucleic acids and proteins, drugs, and nanomaterials [3, 5]. High drug encapsulation is a prime requirement for therapeutic EVs synthesis. Moreover, a major challenge in the encapsulation of drug molecules into the EVs is to maintain the integrity of cell membranes or the bioactivity of the cargo. The strategies of EVs loading, in particular, can be classified into two main classes: endogenous and exogenous loading (Figure 5.2).

5.1.1 Endogenous Loading

Endogenous loading involves modifying the parental cells (usually through transfection or coincubation) to introduce the cargo into the parental cells. After that, they will be packaged into the cytosol or membrane of the EVs through intracellular

Biomedical Applications of Extracellular Vesicles, First Edition.
Edited by Zhenhua Li, Xing-Jie Liang, and Ke Cheng.

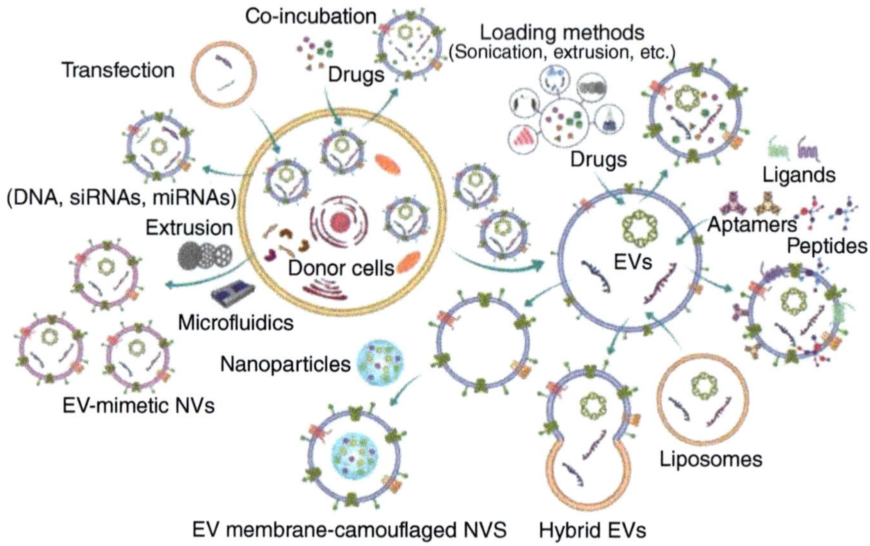

Figure 5.1 Engineering strategies for EVs. Source: Zhang et al. [3]/ John Wiley & Sons.

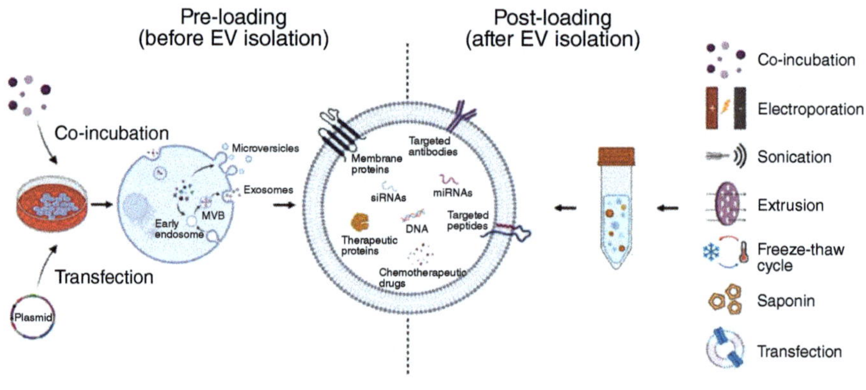

Figure 5.2 Cargo loading strategies for EVs. Source: Wang et al. [6]/with permission of Elsevier.

cargo sorting mechanisms. Transfection is the technique of loading nucleic acids, proteins, and peptides into EVs by introducing plasmid-containing genes of the target molecule into donor cells [7]. After transfection, the cells can ectopically generate desired therapeutic cargos, including miRNAs, siRNAs, and proteins, which are then encapsulated into EVs. This drug loading based on the transfection pattern is simple and highly reproducible, but some limitations of this approach are worth noting such as contamination, poor transfection efficiency, and high dependence on cell viability [8]. miR-132-3p is a well-known angiogenesis miRNA that promotes endothelial cell proliferation, migration, and neovascularization in ischemia hind limb by downregulating RASA1 expression and activating the Ras signaling pathway. Pan et al. showed that miR-132-3p levels in MSC-Exos increased significantly after lentiviral transfection of MSC cells and improved cerebral ischemic injury [9]. Hashimoto et al. constructed engineered exosomes by using a lentiviral infection system to overexpress miR-940 in MDA-MB-231 cells and demonstrated that such exosomes promoted the osteogenic differentiation of human mesenchymal stem cells [10]. The absolute amount of RNAs loaded into the vesicles is related to the size of the RNA molecule and the intracellular concentrations of the transfected nucleic acid. Stoichiometric analyses of EVs showed that only a single copy of small noncoding RNA per several hundred vesicles can be identified, even for highly expressed small, noncoding RNA molecules [11]. HnRNPA2B1 is a ubiquitously expressed RNA-binding protein that controls the transport of a variety of small RNA molecules into exosomes by identifying sequence motifs at the 3′-end of RNAs [12, 13]. To obtain the desired payload of miRNAs, Wang et al. added the exosome-packaging motif to the miRNA401 sequence that can be recognized by HnRNPA2B1. By transfection of plasmid with miRNA401 sequence and exosome-packaging motif into HEp-2 cells, the researchers successfully gained miRNA401-enriched exosomes to reduce virus yields in cells infected by HSV-1 [14]. A variety of such proteins involved in exosome biogenesis have now been identified, including Y-box protein [15] and synaptotagmin-binding cytoplasmic RNA-interacting protein [16]. Fu et al. constructed a hierarchical genetic circuit architecture consisting of a promoter part and a siRNA-expressing part to produce and organize exogenous siRNAs into secretory exosomes. Systemic delivery of EGFR/KRAS siRNAs via these genetic circuits using the liver as a tissue chassis efficiently decreases EGFR/KRAS protein and mRNA in lung cancer [17]. Besides miRNAs and siRNAs, proteins can also be loaded by transfection of the parental cell with the encoding DNA. Aspe et al. obtained Survivin-T34A-contained EVs by transfecting melanoma cells with dominant-negative mutant (Survivin-T34A) [18]. L-Domain is involved in recruiting ESCRT components and ubiquitin ligases to the cell membrane, which play a critical role in MVB formation. Research has shown that the L-domain-containing protein possessed the potential to promote protein loading into exosomes, such as syntenin and Ndfip1. Sterzenbach et al. demonstrated that an intracellular target protein labeled with a WW tag, which can be recognized by the L-domain motifs on Ndfip1, could be loaded into exosomes [19]. An alternative way to enhance drug loading is to utilize a guide molecule that is able to enter exosomes to bring the target cargo into exosomes, such as constitutive proteins or viral proteins. For example, the fusion of target proteins to the exosome-anchoring protein Nef and vesicular stomatitis virus glycoprotein (VSVG) increased their levels in exosomes. As a natural inhibitor of myostatin, myostatin propeptide can enhance muscle growth by inhibiting myostatin; but its efficacy is limited due to low serum stability. Ran et al. improved the delivery efficiency, stability, and treatment efficacy of myostatin propeptide by immobilizing myostatin propeptide to the surface of exosomes through the fusion of the propeptide to an exosomal surface protein CD63 [20]. To improve the loading efficiency of protein drugs, Yim et al. developed a novel strategy for exosomal protein loading based on optically reversible protein–protein interaction called exosomes for protein loading via optically reversible protein–protein interactions

(EXPLORs), which can introduce proteins into exosomes through controlled and reversible loading. The EXPLOR system involves introducing two fusion proteins into the cells: CIBN-conjugated with CD9—an exosome-associated tetraspanin protein and CRY2-conjugated cargo proteins. CIBN and CRY2 can bind to each other after 460–490 nm blue light irradiation, thus inducing a transient docking of CRY2-conjugated cargos to exosomes via endogenous biogenesis [21]. Once the illumination source is removed, cargos can be released into the intraluminal space of the exosome as a result of the separation of CIBN from CRY2 protein [22]. Choi et al. successfully loaded super-repressor IκB (srIκB) into exosomes via the EXPLOR and delivered them to the target cells. Purified srIκB-loaded exosome decreased the NF-κB–mediated inflammatory response, and reduced mortality and systemic inflammation in septic mice [23].

Coincubation is a simple strategy to incubate drugs with parental cells, which enter the cells through interaction with the phospholipid bilayer of the cell membrane, and the drugs are subsequently loaded into the secreted EVs during the process of biogenesis [24, 25]. This strategy is suitable for the loading of hydrophobic drugs, such as doxorubicin and paclitaxel [26–28]. Pascucci et al. found that the membrane microvesicles (MVs) secreted by MSCs after induction with paclitaxel contained paclitaxel and showed antitumor activity [26]. However, the efficiency of loading is usually low and largely influenced by drug properties, drug concentration, and parent cell types [8]. In addition, the drugs may cause damage to parental cells [6].

5.1.2 Exogenous Loading

Exogenous loading is directly loading therapeutic agents into EVs after production and isolation, and includes passive and active loading methods. Passive loading is mainly achieved by simple coincubation EVs with drugs, which can cross the membrane of EVs and passively diffuse into the intraluminal space of EVs. This approach is largely applicable to the loading of hydrophobic drugs, including curcumin, paclitaxel, and doxorubicin [29, 30]. Functionalized nonhydrophobic drug with hydrophobic anchors is an alternative to facilitate the loading efficiency. By conjugating with cholesterol, hydrophilic biologicals siRNA could efficiently load and functionally deliver [31]. The addition of saponin has been proven to enhance the loading efficiency of hydrophilic compounds, even for large proteins [5, 32]. Although this approach can maintain the integrity of the membrane, the loading efficiency is limited and not suitable for larger molecule loading [33].

Active loading strategies require temporary alteration of the permeability of EVs membranes to offer entry of drugs by physical or chemical methods, including electroporation, ultrasound, extrusion, repeated freeze–thawing, with the aid of chemical drugs, transfection, etc. [34, 35]. Electroporation is the preferred technique for loading nucleic acids (such as siRNA, miRNA, and DNA) into EVs. This method typically requires isolation and purification of the EVs before drug loading. Its obvious merits are its simplicity and relatively high drug-loading efficiency. However, the integrity of EVs membrane might be compromised during the loading process.

In addition, an extra step of purification is needed for removing unencapsulated drugs. Electroporation can create transient pores in the EVs membrane by a short, high-voltage pulse in the electric field to promote even larger nucleic acids into exosomes. This loading efficiency appears to be related to the size of the cargos and EVs [36]. Lamichhane et al. showed that the loading efficiency of dsDNA for sizes less than 1000 bp was significantly enhanced compared to larger linear DNAs and plasmid DNAs. They also found that larger EVs can encapsulate more dsDNA than smaller EVs [37]. Moreover, electric field intensity, pulse number, and the type of drug also affect the loading efficiency of electroporation. Although electroporation can improve the loading efficiency of hydrophilic molecules in EVs, one of the most important disadvantages of this technique is the aggregation and fusion of exosomes caused by electroporation [4]. Hood et al. indicated that the addition of trehalose could enhance colloidal stability of the exosomes, and alleviate aggregation and fusion of the EVs [38]. An improved way of loading miRNAs into EVs by using calcium chloride was developed by Zhang et al. This method has shown similar transfection efficacy to electroporation, but it possessed the advantages of easy operation and stability compared to electroporation [39].

Like electroporation, ultrasound enhanced the permeability of the EVs membrane and promoted drug diffusion into EVs during membrane deformation during sonication treatment. Wang et al. successfully loaded the chemotherapeutic drug paclitaxel into M1 macrophages-derived exosomes by slight ultrasound and enhanced their antitumor effect [40]. Compared to coincubation and electroporation, ultrasound loading is more efficient, but the damage to the exosomal membrane caused by ultrasound is not negligible, which may lead to exosome rupture [41].

Cell extrusion generally involves mixing EVs with cargo and squeezing them through a 100–400 nm pore size filter. Exosomes membrane is ruptured and mixed with therapeutic agents during the extrusion to obtain agents' loading, which utilizes the self-organizing properties of the membrane to wrap the cargos [42, 43]. This technique was used to load catalase into naïve Raw 264.7-derived exosomes. Compared with freeze–thaw cycles and coincubation, Haney et al. found that the efficient catalase encapsulation rate by extrusion is the highest, which is similar to ultrasound [32]. However, the membrane structure of EVs may be damaged and cause cargo leakage.

The freeze–thaw cycle needs to mix EVs with cargo, followed by several freezing–thawing cycles in liquid nitrogen or −80 ° C and at room temperature [44]. This method is relatively simple, but its drug-loading capacity is generally lower than that of extrusion and sonication. The size of the EVs may be altered and lead to aggregation. Additionally, proteins consisting in EVs are easily degraded [4, 45].

Direct transfection of agents into isolated EVs using transfection reagents (e.g. liposomes, EXO-Fect, etc.) is also a form of active loading strategy [6]. The loading procedure is quick and easily incorporates cargo into EVs. However, the transfection agents may have a potential toxicity [46].

The application of certain surfactants can alter the permeability of EV membranes, increasing the drug loading capacity of EVs. Saponins are mild surfactants that selectively interact with cholesterol on EVs' membranes, creating transient membrane destabilization that facilitates the entry of drug agents into EVs. Notably, saponins may affect biomolecules, and thus increase toxicity and hemolytic activity [32, 47].

5.2 Engineering EVs for Surface Modification

Although EVs possess a natural targeting ability, the targeting capability of EVs is still fixed and limited in practical applications. Membrane modification is an essential step endowing EVs with additional properties, such as greater targeting ability, and achieving prolonged circulation time and enhanced intracellular uptake [48]. At present, the surface modification strategies include genetic engineering, chemical modification, and membrane fusion.

5.2.1 Genetic Engineering

Genetical engineering is a widely used approach to indirectly modify the exosome-producing cells, which preserves the natural conformation and function of the expressed target protein or peptide by utilizing the inherent protein expression mechanism and the natural biogenesis of the exosome compared with direct modification strategies [49]. Genetic engineering involves fusing a targeting ligand or peptide with a transmembrane protein on the exosome membrane and subsequently transfected with a plasmid encoding the fusion protein into the donor cells for generating surface-modified exosomes.

Lysosome-associated membrane glycoprotein 2b (LAMP-2B) is the most widely used membrane protein for anchoring targeting moieties to the surface of EVs [50]. It has been demonstrated that the N-terminus of LAMP-2B that existed on the surface of EVs can be labeled with targeting sequences [51]. With their properties of small sizes, high affinity and selectivity for their targets, and the lack of immunogenicity, peptides are widely employed as targeting ligands [52]. Xu et al. constructed synovial fluid-derived mesenchymal stem cells (SF-MSCs) targeting exosomes by fusing MSC-binding peptide E7 with LAMP-2B. Compared with exosomes without E7, these E7 functionalized exosomes could effectively deliver the therapeutic drug Kartogenin (KGN) to the SF-MSCs [53]. iRGD (CRGDKGPDC) is a peptide with tumor-targeting potential that can interact with integrin αvβ3 (αvβ3) or αvβ5 and neuropilin-1 (NRP-1) on the tumor cells. By transfecting mouse immature dendritic cells (imDCs) with the pEGFP-C1-RVG- LAMP-2B vector, Tian et al. generated the αv integrin-specific iRGD peptide (CRGDKGPDC)-positive exosomes, which could specifically deliver doxorubicin to αv integrin-positive breast cancer cells [54]. Lin et al. also designed iRGD-engineered exosomes by infecting HEK293T cells with virus iRGD-C1-eGFP-Lamp2b to target delivery of carnitine palmitoyltransferase 1A (CPT1A) siRNAs to colon cancer cells. These iRGD-engineered exosomes showed remarkable tumor targeting and reversed oxaliplatin resistance by delivery of CPT1A siRNAs [55]. Additionally, by transfecting dendritic cells with a plasmid encoding a neuron-specific rabies virus glycoprotein (RVG) fused to LAMP-2B, Alvarez-Erviti et al. constructed a brain-targeting exosome [50]. Despite the effectiveness of anchoring peptides into the surface of EVs, in some cases, the peptides fused to the N terminus of LAMP-2B can be degraded by endosomal proteases and cannot be effectively displayed on the exosome surface. Leonard et al. inserted glycosylation sequence (GNSTM) at the N-terminal end of peptide-LAMP-2B fusion protein to prevent the degradation of the peptide [56].

Except for peptides, proteins or antibody fragments can also be genetically fused with LAMP-2B to be presented on the surface of exosomes. Bellavia et al. used the IL3 as a targeting ligand in the myeloid leukemia (CML), because the IL3 receptor (IL3-R) is overexpressed in CML cells compared to normal hematopoietic cells. By fusing IL-3 to the N-terminal of LAMP-2B, the IL-3 is successfully present on the surface of exosomes. Guided by IL-3, imatinib and BCR-ABL siRNA encapsulated in this engineered exosome could be efficiently delivered to tumor tissue and kill CML cells [57]. Si et al. generated T cells-targeted exosomes containing anti-CD3/CD28 single-chain variable fragments (scFvs) expressed on the surface of exosomes by fusing it to the N-terminus of LAMP-2B. Researchers demonstrated that CD3/CD28 ScFv expressed on exosomes' membranes could bind to T cells and showed excellent cancer cell-killing capacity by delivering CAR mRNA and anti-CD3/CD28 scFvs to T cells [58]. Cavin-2 plays a major role in regulating EVs uptake. Liao et al. produced Cavin-2-LAMP-2B engineered exosomes by transfecting MSCs with GV492-CAVIN2-LAMP-2B vector, which promoted the uptake of EVs by nucleus pulposus cells (NPCs) under TNF-α treatment [59].

In addition to LAMP-2B, other membrane proteins expressed on exosomes such as CD63, CD9, CD81, GPI, PDGFRs, etc. are also used to fuse with targeting molecules and display them on the surface of exosomes [20, 52, 60, 61]. By transfecting donor cells with vector encoding PDGFR fused to the GE11 peptide, Ohno et al. generated epidermal growth factor receptor (EGFR)-targeted exosomes. By intravenous injection, these exosomes delivered let-7a miRNA to EGFR-expressing breast cancer tissue [7]. A single polypeptide consisting of two scFv antibodies targeting CD3 and EGFR was genetically fused to PDGFR for the generation of synthetic multivalent antibodies retargeted exosomes (SMART-Exos) in Expi293F cells. SMART-Exos showed high affinity for T-cell CD3 and cancer cell-associated EGFR, making them conjugate T cells to EGFR-expressing breast cancer cells, and elicit potent antitumor immunity [62]. A targeting sequence or probe can also be engineered into the tetraspanin superfamily CD63/CD9/CD81. As a known target of the scavenger receptor class B type 1 (SR-B1) receptor, Apo-A1 is employed to be fused with the transmembrane protein CD63 for producing Apo-A1-CD63-presenting exosomes. These engineered exosomes were able to specifically deliver miR-26a to tumor cells with high SR-B1 expression [63]. Furthermore, Kanuma et al. genetically inserted the gene encoding ovalbumin (OVA) antigen into the extracellular loop of CD63 to generate OVA-carrying EVs. Injection of OVA-carrying EVs induced potent Ag-specific T-cell responses [64]. Lactadherin (an exosome-tropic protein)-bound antigen agents can also be displayed on the surface of EVs through the C1C2 structural domain of lactadherin, which binds to exosome lipids [21, 65, 66]. While genetically modification of exosome-producing cells can load targeted cargo on the surface of EVs, the genetic engineering strategy has several challenges. For example, this method is restricted to targeting cargos that can be genetically encoded. Moreover, the introduction of the targeting moiety may affect the functional profile of exosome membrane proteins. Notably, exposure of peptide- LAMP-2B fusion protein to endosomal proteases during EVs biogenesis may result in peptide loss. Although this phenomenon can be overcome by introducing a stabilizing glycosylation motif, there is a need to closely monitor the integrity of the targeting, pharmaceutical, or labeling moiety of interest when employing genetic engineering strategies.

5.2.2 Chemical Modification

The chemical modification relies on the biological coupling between targeted ligands and surface proteins of EVs. It can be divided into noncovalent modification and covalent modification.

Among them, noncovalent modification mainly depends on hydrophobic interaction and receptor–ligand interaction. Modification that depends on hydrophobic interaction needs to connect a hydrophobic tail on the target molecule and modify the target molecules on the EVs surface through the hydrophobic interaction between the tail and the EVs membrane. For example, Huang et al. connected the tail of polyethylene glycol–ethanolamine phosphate (PEG–DMPE) on the cyclic peptide c (RGDyK), so that the cyclic peptide c (RGDyK) is anchored on the EVs surface, making them target osteosarcoma cells [67]. Connected the cholesterol tail on the nucleic acid aptamer of prostate-specific membrane antigen, Pi et al. manufactured a prostate cancer cell-targeted EVs via the surface-modified nucleic acid aptamer on the surface of EVs. The RNA ligand-displaying EVs were specifically localized to prostate cancer tissues, making them a potential delivery system for prostate cancer cells. In mice, the PSMA aptamer displaying EVs specifically delivered survivin siRNA to prostate cancer tissues and effectively inhibited prostate cancer xenograft growth [68]. Receptor–ligand interaction can also modify the target molecules on the surface of EVs. Reticulocyte-derived EVs have transferrin receptors. Qi et al. connected superparamagnetic nanoparticles with transferrin and modified superparamagnetic nanoparticles on the surface of EV. Depending on the interaction between transferrin and transferrin receptors, the EVs can target tumor sites under the guidance of a magnetic field [69].

Covalent modification requires the formation of a covalent bond between the modified group and the EVs membrane protein, which usually requires harsh conditions such as high temperature or heavy metal solvent [70]. In recent years, researchers have developed some covalent modification methods with relatively mild conditions. Pham et al. found that EGFR-targeted peptides can be connected to EVs membrane proteins through peptide bonds by using efficient protein ligases, such as Sortase A and OaAEP1 ligases [71]. Chemical engineering with simple molecules can be carried out in one step through the reaction with amino acid side chains of EVs surface proteins. The most abundant suitable functional groups on the EVs surface are the amino groups available in lysine side chains and the

N-termini of EVs surface proteins. They can be easily addressed using activated esters, e.g. *N*-hydroxysuccinimide esters (NHS esters). Amino groups can be used directly as an anchor point for chemically simple molecules, such as fluorescent dyes [72] or PEG derivatives [73].

More complex molecules can be coupled chemically to EVs by first introducing azide groups. Researchers have mainly focused on azide alkyne cycloaddition reactions to date. These reactions can be divided into copper-catalyzed alkyne-azide cycloaddition (CuAAC) and its strain-promoted counterpart (SPAAC). Smyth et al. introduced pentanoic acid to vesicle surface proteins and subsequently coupled an azide-functionalized fluorescent dye via CuAAC [74]. CuAAC can reach higher rate constants, and the simple alkynes needed in these reactions are chemically more accessible and less bulky than the strained cyclooctynes employed in SPAAC. However, the latter approach does not require the addition of Cu[I]-ions. These ions might cause oxidative damage to EV surface proteins and need to be thoroughly removed from the vesicles before EV application *in vitro* and *in vivo* due to their toxicity. To what extent copper ions might damage EVs has not been explored yet. SPAAC needs to introduce the reaction group into the surface of EV first [75]. For example, Zhu et al. crosslinked dibenzobicyclooctyne (DBCO) to the surface of dendritic cell EVs through a chemical reaction and then made it react with azidoated tumor antigen glycoprotein MUC1, thereby covalently connecting MUC1 to the surface of dendritic cell EVs [76]. SPAAC between dibenzocyclooctyne (DBCO) on the surface of EVs and an azide-labeled cRGD-peptide was employed to produce EVs that selectively targeted ischemic lesions in the brain of mice [77].

Compared with genetic modification, chemical modification can modify more diverse molecules on the surface of EV, such as nucleic acid, lipids, nanoparticles, etc. While the basics of chemical surface engineering of EVs have been explored, there are still opportunities to further optimize its application. The mild reaction conditions and high selectivity of SPAAC that have led to its wide application, even for direct labeling *in vivo*, predestine it as the reaction of choice for EV labeling [78]. With its easy implementation, chemical modification of EVs has great potential as a standardized method that can be readily applied to endow EVs derived from different sources with a wide variety of targeting moieties, fluorescent dyes, or PEG derivatives. However, to reach this goal the effects of chemically linking molecules to the amino groups of functional EV surface proteins warrant further evaluation, as these modifications might affect the natural activity of EVs to an unknown extent. In addition, the presence of nonvesicular material carried over from the medium after ultracentrifugation or PEG-precipitation can lead to unwanted side reactions that may interfere with the surface modification of the vesicles [79]. Thus, during the development and optimization of chemical surface engineering, the purity of EV-samples needs to be taken into account.

5.2.3 Hydrophobic Membrane Engineering

Combining the advantages of both liposomes and exosomes, membrane fusion-based hybrid exosomes (MFHEs) have emerged as novel nanoparticles for drug delivery and are generated by fusing exosomes with liposomes via different membrane fusion methods [80]. MFHEs not only have the characteristics of high drug loading rate, stability, and easy surface modification, but they are also endowed with high biocompatibility and low immunogenicity of exosomes. MFHEs are promising to be applied to cancer therapy and other disease treatments as well as disease diagnosis. Membrane fusion can be achieved mainly through the following means, including freeze–thaw, natural incubation, polyethylene glycol-mediated fusion, and membrane extrusion method.

Freeze–thaw method is commonly used for loading drugs to liposomes. By temporarily forming ice crystals, the plasma membrane can be disrupted and water-soluble molecules can be loaded into the liposomes [81]. Sato et al. mixed Raw 264.7 cell-derived exosomes with dual fluorescently labeled liposomes (1: 1 by volume). The mixture was then frozen in liquid nitrogen and thawed subsequently at room temperature for 15 minutes. After repeated freeze–thaw cycles, the resulting hybrid exosomes showed a higher cellular uptake rate compared with liposomes [45]. Cheng et al. designed hybrid exosomes by fusing genetically engineered exosomes with heat-sensitive liposomes to combine photothermal therapy with immunotherapy for cancer treatment [82]. Although freeze–thaw method may achieve high fusion efficiency, it also has significant downsides. High-frequency freeze–thaw cycles may not only affect the biological activity of pharmaceuticals in the vesicles but may also destroy the biological integrity of the exosome membrane. When choosing this fusion method, the types of drugs and types of freeze–thaw cycles need to be taken into careful consideration.

Hybrid exosomes are formed through electrostatic or hydrophobic interactions without disrupting the integrity of the lipid bilayer or leaking the vesicle contents. Lin et al. constructed hybrid exosomes by simply incubating HEK293FT cell-derived exosomes with CRISPR/Cas9-expressing liposomes at 37 °C for 12 hours and provided a new idea for the safe and effective delivery of the CRISPR-Cas9 system [83]. Although this method does little damage to vesicles and drugs, the fusion efficiency is relatively low.

Polyethylene glycol (PEG) can change cell membranes and has been widely used to mediate cell-to-cell fusion by mediating the close contact of lipid bilayer membrane structures, and triggering the evacuation and reorganization of lipid molecules [84]. Piffoux et al. demonstrated that PEG could induce the fusion of exosomes and liposomes from different cell sources [85]. Since PEG is easy to prepare and stable in activity, it can not only mediate the efficient fusion of exosomes and liposomes but also increase their circulation time in the blood. However, it may reduce the cellular uptake of the hybrid exosomes [86].

The membrane extrusion method refers to the simultaneous extrusion of exosomes and liposomes through membrane pores with a controllable size under physical pressure to form mixed vesicles. Sun et al. designed hybrid nanovesicles using clodronate-loaded (CLD) liposomes and fibroblast-derived exosomes for the treatment of pulmonary fibrosis. The researchers mixed L-929 fibroblast-derived exosomes with a suspension of synthetic liposomes at a 1: 5 protein equivalent ratio, vortexed and sonicated the mixture, and then passed it through 400 and 200 nm polycarbonate ester films with repeated extrusion (10 times). This resulted in the successful formation of exosome-hybridized liposomes [87]. Both the pore size of the polycarbonate membrane and the frequency of membrane extrusion impact the features of MFHEs. Although membrane extrusion methods show high fusion efficiency, the shear stress generated during the extrusion process may damage the membrane integrity of natural exosomes.

References

1 Wiklander, O.P.B., Brennan, M., Lötvall, J. et al. (2019). Advances in therapeutic applications of extracellular vesicles. Science translational medicine. *Science* 11 (492): eaav8521.

2 Théry, C., Witwer, K.W., Aikawa, E. et al. (2018). Minimal information for studies of extracellular vesicles 2018 (MISEV2018): a position statement of the International Society for Extracellular Vesicles and update of the MISEV2014 guidelines. *J. Extracell. Vesicles* 7 (1): 1535750.

3 Zhang, X., Zhang, H., Gu, J. et al. (2021). Engineered extracellular vesicles for cancer therapy. *Adv. Mater.* 33 (14): e2005709.

4 Zhang, Y., Bi, J., Huang, J. et al. (2020). Exosome: a review of its classification, isolation techniques, storage, diagnostic and targeted therapy applications. *Int. J. Nanomed.* 15: 6917–6934.

5 Herrmann, I.K., Wood, M.J.A., and Fuhrmann, G. (2021). Extracellular vesicles as a next-generation drug delivery platform. *Nat. Nanotechnol.* 16 (7): 748–759.

6 Wang, Z., Mo, H., He, Z. et al. (2022). Extracellular vesicles as an emerging drug delivery system for cancer treatment: current strategies and recent advances. *Biomed. Pharmacother.* 153: 113480.

7 Ohno, S., Takanashi, M., Sudo, K. et al. (2013). Systemically injected exosomes targeted to EGFR deliver antitumor microRNA to breast cancer cells. *Mol. Ther.* 21 (1): 185–191.

8 Chavda, V.P., Pandya, A., Kumar, L. et al. (2023). Exosome nanovesicles: a potential carrier for therapeutic delivery. *Nano Today* 49: 101771.

9 Pan, Q., Kuang, X., Cai, S. et al. (2020). miR-132-3p priming enhances the effects of mesenchymal stromal cell-derived exosomes on ameliorating brain ischemic injury. *Stem Cell Res. Ther.* 11 (1): 260.

10 Hashimoto, K., Ochi, H., Sunamura, S. et al. (2018). Cancer-secreted hsa-miR-940 induces an osteoblastic phenotype in the bone metastatic microenvironment via targeting ARHGAP1 and FAM134A. *Proc. Natl. Acad. Sci. U.S.A.* 115 (9): 2204–2209.

11 Buzas, E.I. (2022). The roles of extracellular vesicles in the immune system. *Nat. Rev. Immunol.* 1–15.

12 Villarroya-Beltri, C., Gutiérrez-Vázquez, C., Sánchez-Cabo, F. et al. (2013). Sumoylated hnRNPA2B1 controls the sorting of miRNAs into exosomes through binding to specific motifs. *Nat. Commun.* 4: 2980.

13 Zhou, X., Wang, L., Zou, W. et al. (2020). hnRNPA2B1 associated with recruitment of RNA into exosomes plays a key role in herpes simplex virus 1 release from infected cells. *J. Virol.* 94 (13): 10–1128.

14 Wang, L., Chen, X., Zhou, X. et al. (2018). miRNAs targeting ICP4 and delivered to susceptible cells in exosomes block HSV-1 replication in a dose-dependent manner. *Mol. Ther.* 26 (4): 1032–1039.

15 Shurtleff, M.J., Temoche-Diaz, M.M., Karfilis, K.V. et al. (2016). Y-box protein 1 is required to sort microRNAs into exosomes in cells and in a cell-free reaction. *Elife* 5.

16 Santangelo, L., Giurato, G., Cicchini, C. et al. (2016). The RNA-binding protein SYNCRIP is a component of the hepatocyte exosomal machinery controlling microRNA sorting. *Cell Rep.* 17 (3): 799–808.

17 Fu, Z., Zhang, X., Zhou, X. et al. (2021). In vivo self-assembled small RNAs as a new generation of RNAi therapeutics. *Cell Res.* 31 (6): 631–648.

18 Aspe, J.R., Diaz Osterman, C.J., Jutzy, J.M. et al. (2014). Enhancement of gemcitabine sensitivity in pancreatic adenocarcinoma by novel exosome-mediated delivery of the Survivin-T34A mutant. *J. Extracell. Vesicles* 3.

19 Sterzenbach, U., Putz, U., Low, L.H. et al. (2017). Engineered Exosomes as Vehicles for Biologically Active Proteins. *Mol Ther.* 25 (6): 1269–1278.

20 Ran, N., Gao, X., Dong, X. et al. (2020). Effects of exosome-mediated delivery of myostatin propeptide on functional recovery of mdx mice. *Biomaterials* 236: 119826.

21 Alptekin, A., Parvin, M., Chowdhury, H.I. et al. (2022). Engineered exosomes for studies in tumor immunology. *Immunol. Rev.* 312 (1): 76–102.

22 Yim, N., Ryu, S.W., Choi, K. et al. (2016). Exosome engineering for efficient intracellular delivery of soluble proteins using optically reversible protein-protein interaction module. *Nat. Commun.* 7: 12277.

23 Choi, H., Kim, Y., Mirzaaghasi, A. et al. (2020). Exosome-based delivery of super-repressor IκBα relieves sepsis-associated organ damage and mortality. *Sci. Adv.* 6 (15): eaaz6980.

24 Wunsch, B.H., Smith, J.T., Gifford, S.M. et al. (2016). Nanoscale lateral displacement arrays for the separation of exosomes and colloids down to 20 nm. *Nat. Nanotechnol.* 11 (11): 936–940.

25 Ahmadi, M., Hassanpour, M., and Rezaie, J. (2022). Engineered extracellular vesicles: a novel platform for cancer combination therapy and cancer immunotherapy. *Life Sci.* 308: 120935.

26 Pascucci, L., Coccè, V., Bonomi, A. et al. (2014). Paclitaxel is incorporated by mesenchymal stromal cells and released in exosomes that inhibit in vitro tumor growth: a new approach for drug delivery. *J. Controlled Release* 192: 262–270.

27 Bonomi, A., Silini, A., Vertua, E. et al. (2015). Human amniotic mesenchymal stromal cells (hAMSCs) as potential vehicles for drug delivery in cancer therapy: an in vitro study. *Stem Cell Res. Ther.* 6 (1): 155.

28 Yang, Y., Chen, Y., Zhang, F. et al. (2015). Increased anti-tumour activity by exosomes derived from doxorubicin-treated tumour cells via heat stress. *Int. J. Hyperthermia* 31 (5): 498–506.

29 Yang, T., Martin, P., Fogarty, B. et al. (2015). Exosome delivered anticancer drugs across the blood-brain barrier for brain cancer therapy in Danio rerio. *Pharm. Res.* 32 (6): 2003–2014.

30 Sun, D., Zhuang, X., Xiang, X. et al. (2010). A novel nanoparticle drug delivery system: the anti-inflammatory activity of curcumin is enhanced when encapsulated in exosomes. *Mol. Ther.* 18 (9): 1606–1614.

31 O'Loughlin, A.J., Mäger, I., de Jong, O.G. et al. (2017). Functional delivery of lipid-conjugated siRNA by extracellular vesicles. *Mol. Ther.* 25 (7): 1580–1587.

32 Haney, M.J., Klyachko, N.L., Zhao, Y. et al. (2015). Exosomes as drug delivery vehicles for Parkinson's disease therapy. *J. Controlled Release* 207: 18–30.

33 de Jong, O.G., Kooijmans, S.A.A., Murphy, D.E. et al. (2019). Drug delivery with extracellular vesicles: from imagination to innovation. *Acc Chem Res.* 52 (7): 1761–1770.

34 Liu, C., Wang, Y., Li, L. et al. (2022). Engineered extracellular vesicles and their mimetics for cancer immunotherapy. *J. Controlled Release* 349: 679–698.

35 Fuhrmann, G., Serio, A., Mazo, M. et al. (2015). Active loading into extracellular vesicles significantly improves the cellular uptake and photodynamic effect of porphyrins. *J. Controlled Release* 205: 35–44.

36 Zhang, Y.F., Shi, J.B., and Li, C. (2019). Small extracellular vesicle loading systems in cancer therapy: current status and the way forward. *Cytotherapy* 21 (11): 1122–1136.

37 Lamichhane, T.N., Raiker, R.S., and Jay, S.M. (2015). Exogenous DNA loading into extracellular vesicles via electroporation is size-dependent and enables limited gene delivery. *Mol. Pharm.* 12 (10): 3650–3657.

38 Hood, J.L., Scott, M.J., and Wickline, S.A. (2014). Maximizing exosome colloidal stability following electroporation. *Anal. Biochem.* 448: 41–49.

39 Zhang, D., Lee, H., Zhu, Z. et al. (2017). Enrichment of selective miRNAs in exosomes and delivery of exosomal miRNAs in vitro and in vivo. *Am. J. Physiol. Lung Cell. Mol. Physiol.* 312 (1): L110–l21.

40 Wang, P., Wang, H., Huang, Q. et al. (2019). Exosomes from M1-polarized macrophages enhance paclitaxel antitumor activity by activating macrophages-mediated inflammation. *Theranostics* 9 (6): 1714–1727.

41 Kim, M.S., Haney, M.J., Zhao, Y. et al. (2016). Development of exosome-encapsulated paclitaxel to overcome MDR in cancer cells. *Nanomed. Nanotechnol. Biol. Med.* 12 (3): 655–664.

42 Goh, W.J., Zou, S., Ong, W.Y. et al. (2017). Bioinspired cell-derived nanovesicles versus exosomes as drug delivery systems: a cost-effective alternative. *Sci. Rep.* 7 (1): 14322.

43 Park, K.S., Svennerholm, K., Shelke, G.V. et al. (2019). Mesenchymal stromal cell-derived nanovesicles ameliorate bacterial outer membrane vesicle-induced sepsis via IL-10. *Stem Cell Res. Ther.* 10 (1): 231.

44 Luan, X., Sansanaphongpricha, K., Myers, I. et al. (2017). Engineering exosomes as refined biological nanoplatforms for drug delivery. *Acta Pharm. Sin.* 38 (6): 754–763.

45 Sato, Y.T., Umezaki, K., Sawada, S. et al. (2016). Engineering hybrid exosomes by membrane fusion with liposomes. *Sci. Rep.* 6: 21933.

46 Wang, T., Larcher, L.M., Ma, L., and Veedu, R.N. (2018). Systematic screening of commonly used commercial transfection reagents towards efficient transfection of single-stranded oligonucleotides. *Molecules (Basel, Switzerland)* 23 (10): 2564.

47 Podolak, I., Galanty, A., and Sobolewska, D. (2010). Saponins as cytotoxic agents: a review. *Phytochem. Rev.* 9 (3): 425–474.

48 Tang, T.T., Wang, B., Lv, L.L., and Liu, B.C. (2020). Extracellular vesicle-based nanotherapeutics: emerging frontiers in anti-inflammatory therapy. *Theranostics* 10 (18): 8111–8129.

49 Liang, Y., Duan, L., Lu, J., and Xia, J. (2021). Engineering exosomes for targeted drug delivery. *Theranostics* 11 (7): 3183–3195.

50 Alvarez-Erviti, L., Seow, Y., Yin, H. et al. (2011). Delivery of siRNA to the mouse brain by systemic injection of targeted exosomes. *Nat. Biotechnol.* 29 (4): 341–345.

51 Li, Z., Zhou, X., Gao, X. et al. (2020). Fusion protein engineered exosomes for targeted degradation of specific RNAs in lysosomes: a proof-of-concept study. *J. Extracell. Vesicles* 9 (1): 1816710.

52 Salunkhe, S., Basak, M., Chitkara, D., and Mittal, A. (2020). Surface functionalization of exosomes for target-specific delivery and in vivo imaging & tracking: Strategies and significance. *J. Controlled Release* 326: 599–614.

53 Xu, X., Liang, Y., Li, X. et al. (2021). Exosome-mediated delivery of kartogenin for chondrogenesis of synovial fluid-derived mesenchymal stem cells and cartilage regeneration. *Biomaterials* 269: 120539.

54 Tian, Y., Li, S., Song, J. et al. (2014). A doxorubicin delivery platform using engineered natural membrane vesicle exosomes for targeted tumor therapy. *Biomaterials* 35 (7): 2383–2390.

55 Lin, D., Zhang, H., Liu, R. et al. (2021). iRGD-modified exosomes effectively deliver CPT1A siRNA to colon cancer cells, reversing oxaliplatin resistance by regulating fatty acid oxidation. *Mol. Oncol.* 15 (12): 3430–3446.

56 Hung, M.E. and Leonard, J.N. (2015). Stabilization of exosome-targeting peptides via engineered glycosylation. *J. Biol. Chem.* 290 (13): 8166–8172.

57 Bellavia, D., Raimondo, S., Calabrese, G. et al. (2017). Interleukin 3- receptor targeted exosomes inhibit in vitro and in vivo Chronic Myelogenous Leukemia cell growth. *Theranostics* 7 (5): 1333–1345.

58 Si, K., Dai, Z., Li, Z. et al. (2023). Engineered exosome-mediated messenger RNA and single-chain variable fragment delivery for human chimeric antigen receptor T-cell engineering. *Cytotherapy* 6: 615–624.

59 Liao, Z., Liu, H., Ma, L. et al. (2021). Engineering extracellular vesicles restore the impaired cellular uptake and attenuate intervertebral disc degeneration. *ACS Nano.* 15 (9): 14709–14724.

60 Li, Z., Zhou, X., Wei, M. et al. (2019). In vitro and in vivo RNA inhibition by CD9-HuR functionalized exosomes encapsulated with miRNA or CRISPR/dCas9. *Nano Lett.* 19 (1): 19–28.

61 Liu, W., Huang, Y., Li, Z. et al. (2022). Multivalent engineering of exosomes with activatable aptamer probes for specific regulation and monitoring of cell targeting. *Anal. Chem.* 94 (9): 3840–3848.

62 Cheng, Q., Shi, X., Han, M. et al. (2018). Reprogramming exosomes as nanoscale controllers of cellular immunity. *J. Am. Chem. Soci.* 140 (48): 16413–16417.

63 Liang, G., Kan, S., Zhu, Y. et al. (2018). Engineered exosome-mediated delivery of functionally active miR-26a and its enhanced suppression effect in HepG2 cells. *Int. J. Nanomed.* 13: 585–599.

64 Kanuma, T., Yamamoto, T., Kobiyama, K. et al. (2017). CD63-mediated antigen delivery into extracellular vesicles via DNA vaccination results in robust CD8(+) T cell responses. *J. Immunol. (Baltimore, MD)* 198 (12): 4707–4715.

65 Imai, T., Takahashi, Y., Nishikawa, M. et al. (2015). Macrophage-dependent clearance of systemically administered B16BL6-derived exosomes from the blood circulation in mice. *J. Extracell. Vesicles* 4: 26238.

66 Takahashi, Y., Nishikawa, M., Shinotsuka, H. et al. (2013). Visualization and in vivo tracking of the exosomes of murine melanoma B16-BL6 cells in mice after intravenous injection. *J. Biotechnol.* 165 (2): 77–84.

67 Huang, X., Wu, W., Jing, D. et al. (2022). Engineered exosome as targeted lncRNA MEG3 delivery vehicles for osteosarcoma therapy. *J. Controlled Release* 343: 107–117.

68 Pi, F., Binzel, D.W., Lee, T.J. et al. (2018). Nanoparticle orientation to control RNA loading and ligand display on extracellular vesicles for cancer regression. *Nat. Nanotechnol.* 13 (1): 82–89.

69 Qi, H., Liu, C., Long, L. et al. (2016). Blood exosomes endowed with magnetic and targeting properties for cancer therapy. *ACS Nano* 10 (3): 3323–3333.

70 Armstrong, J.P., Holme, M.N., and Stevens, M.M. (2017). Re-engineering extracellular vesicles as smart nanoscale therapeutics. *ACS Nano* 11 (1): 69–83.

71 Pham, T.C., Jayasinghe, M.K., Pham, T.T. et al. (2021). Covalent conjugation of extracellular vesicles with peptides and nanobodies for targeted therapeutic delivery. *J. Extracell. Vesicles* 10 (4): e12057.

72 Kooijmans, S.A.A., Gitz-Francois, J., Schiffelers, R.M., and Vader, P. (2018). Recombinant phosphatidylserine-binding nanobodies for targeting of extracellular vesicles to tumor cells: a plug-and-play approach. *Nanoscale* 10 (5): 2413–2426.

73 Choi, E.S., Song, J., Kang, Y.Y., and Mok, H. (2019). Mannose-modified serum exosomes for the elevated uptake to murine dendritic cells and lymphatic accumulation. *Macromol. Biosci.* 19 (7): e1900042.

74 Smyth, T., Petrova, K., Payton, N.M. et al. (2014). Surface functionalization of exosomes using click chemistry. *Bioconjugate Chem.* 25 (10): 1777–1784.

75 Richardson, J.J. and Ejima, H. (2019). Surface engineering of extracellular vesicles through chemical and biological strategies. *Chem. Mater.* 31 (7): 2191–2201.

76 Zhu, H., Wang, K., Wang, Z. et al. (2022). An efficient and safe MUC1-dendritic cell-derived exosome conjugate vaccine elicits potent cellular and humoral immunity and tumor inhibition in vivo. *Acta Biomater.* 138: 491–504.

77 Tian, T., Zhang, H.X., He, C.P. et al. (2018). Surface functionalized exosomes as targeted drug delivery vehicles for cerebral ischemia therapy. *Biomaterials* 150: 137–149.
78 Chang, P.V., Prescher, J.A., Sletten, E.M. et al. (2010). Copper-free click chemistry in living animals. *Proc. Natl. Acad. Sci. U.S.A.* 107 (5): 1821–1826.
79 Li, P., Kaslan, M., Lee, S.H. et al. (2017). Progress in exosome isolation techniques. *Theranostics* 7 (3): 789–804.
80 Lu, M. and Huang, Y. (2020). Bioinspired exosome-like therapeutics and delivery nanoplatforms. *Biomaterials* 242: 119925.
81 Costa, A.P., Xu, X., and Burgess, D.J. (2014). Freeze-anneal-thaw cycling of unilamellar liposomes: effect on encapsulation efficiency. *Pharm. Res.* 31 (1): 97–103.
82 Cheng, L., Zhang, X., Tang, J. et al. (2021). Gene-engineered exosomes-thermosensitive liposomes hybrid nanovesicles by the blockade of CD47 signal for combined photothermal therapy and cancer immunotherapy. *Biomaterials* 275: 120964.
83 Lin, Y., Wu, J., Gu, W. et al. (2018). Exosome-liposome hybrid nanoparticles deliver CRISPR/Cas9 system in MSCs. *Adv. Sci.* 5 (4): 1700611.
84 Lentz, B.R. (2007). PEG as a tool to gain insight into membrane fusion. *Eur. Biophys. J.* 36 (4–5): 315–326.
85 Piffoux, M., Silva, A.K.A., Wilhelm, C. et al. (2018). Modification of extracellular vesicles by fusion with liposomes for the design of personalized biogenic drug delivery systems. *ACS Nano* 12 (7): 6830–6842.
86 Kannavou, M., Marazioti, A., Stathopoulos, G.T., and Antimisiaris, S.G. (2021). Engineered versus hybrid cellular vesicles as efficient drug delivery systems: a comparative study with brain targeted vesicles. *Drug Delivery Transl. Res.* 11 (2): 547–565.
87 Sun, L., Fan, M., Huang, D. et al. (2021). Clodronate-loaded liposomal and fibroblast-derived exosomal hybrid system for enhanced drug delivery to pulmonary fibrosis. *Biomaterials* 271: 120761.

6

Current Technology for Production, Isolation, and Quality Control of Extracellular Vesicles

Dandan Han[1], *Yichuan Ma*[1], *Yujing Hu*[1], *and Zhenhua Li*[2,3]

[1] *Hebei University, College of Chemistry & Material Science, Key Laboratory of Medicinal Chemistry and Molecular Diagnosis of Ministry of Education, Chemical Biology Key Laboratory of Hebei Province, Wusi East Avenue, Baoding Hebei 071002, China*
[2] *The Tenth Affiliated Hospital of Southern Medical University, Dongguan, Guangdong 523059, China*
[3] *Guangdong Provincial Key Laboratory of Cardiac Function and Microcirculation, Guangzhou 510515, China*

Extracellular vesicles (EVs), derived from cell membranes, exhibit tremendous potential as therapeutic agents and drug carriers. Despite their promising applications, the development of high-quality and scalable EVs for clinical translation poses significant challenges. The isolation, enrichment, and separation of EV samples are continue to face obstacles. Further research is essential to refine these processes and fully comprehend the intrinsic properties of EVs, thereby confirming their enormous potential for clinical use.

6.1 Production of EVs

The secretion of EVs occurs naturally, but the process is slow, and the structures and compositions of EVs are highly diverse. Consequently, translating EVs into clinical applications can be challenging, necessitating alternative approaches for EV generation. The following outlines five primary methods: (i) three-dimensional culture; (ii) physical stimulation; (iii) chemical stimulation; (iv) physiological modification; and (v) genetic manipulation.

6.1.1 Three-Dimensional Culture

One way to enhance the secretion of EVs in cultured cells is to utilize a 3D cell culture system, such as a bioreactor or cell ball. Such systems allow for a high-density expansion of cells, resulting in a high yield of EVs due to the large surface area provided for cell growth (as illustrated in Figure 6.1). The bioreactor is a commonly used 3D cell culture system for mass production of EVs that not only increases

Biomedical Applications of Extracellular Vesicles, First Edition.
Edited by Zhenhua Li, Xing-Jie Liang, and Ke Cheng.

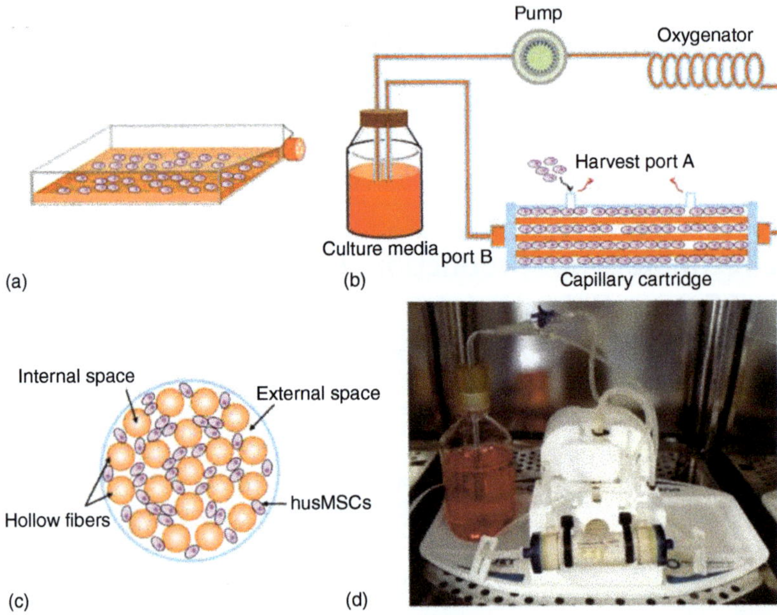

Figure 6.1 Schematic and photograph of the 2D and 3D culture system. Source: Adapted from Dionysios et al. [1]. (a) The schematic of conventional 2D flask. (b) The general schematic of the hollow fiber bioreactor-based 3D culture system. The system was composed of a pulsatile perfusion pump, an oxygenator, a cartridge containing thousands of hollow fibers, a bottle of culture medium, and the connecting tube. (c) A general schematic of cross-sectional view through the bioreactor. (d) The photograph of our 3D culture system.

yield but also provides mechanical stimulation to stimulate cells and enhance EV production [2]. Using a bioreactor can improve yield and concentration of EVs in the culture medium, as well as shorten production time. One study demonstrated that umbilical cord-derived mesenchymal stem cells (UC-MSCs) cultured in a 3D hollow fiber bioreactor secreted 7.5 times more EVs than cells cultured in a 2D tissue culture flask. Furthermore, 3D-EVs were more effective in promoting chondrocyte proliferation and migration, and inhibiting chondrocyte apoptosis *in vitro*. The study also showed that 3D-EVs were more effective in promoting cartilage regeneration in a cartilage defect model *in vivo* [3]. Cao et al. found that the protein yield of UC-MSCs small EVs cultured in a 3D hollow fiber bioreactor was 19.4 times higher than that of cells cultured in 2D. *In vivo* experimental results showed that 3D-EVs were more effective than 2D-EVs in improving acute renal injury induced by cisplatin, which showed better renal function, less-severe renal tubular lesions, and less inflammatory cell infiltration. The authors attributed the renal protective effect of 3D-EVs to the increase of their absorption by tubular epithelial cells and the enhancement of anti-inflammatory activity [4]. Watson et al. reported that when human embryonic kidney (HEK) 293 cells stably expressing hetIL-15 were cultured in a hollow fiber bioreactor, the secretion of small EVs increased by 40 times compared to standard 2D culture [1]. The study also found that 3D-EVs contained less serum protein pollutants than 2D-EVs. Other studies have reported large-scale production of EVs using bioreactors [5], demonstrating the feasibility of mass-producing cell-derived EVs using this technology. In addition to bioreactors, 3D spherical culture has also been shown to increase the secretion of EVs. For example, EVs produced by 3D hanging drop spheres and 3D poly (2-2-hydroxyethyl methacrylate) spheres were significantly higher than those produced by 2D cultured bone mesenchymal stem cells (BM-MSCs). Interestingly, the study found that increasing the size of 3D spheres resulted in a decrease in EV secretion [6]. Therefore, when utilizing 3D spherical culture for mass production of EVs, size may be an important parameter to consider for optimization.

6.1.2 Physical Stimulation

Various physical stimulation techniques, including irradiation, electric and magnetic field stimulation, mechanical stimulation, and terrain clues, have been researched and utilized to enhance the production of EVs. Generally, physical stimulation induces stress in cells, prompting them to generate more EVs, as depicted in Figure 6.2 [8]. Ionizing radiation, for instance, has been observed to increase the number of EVs produced by cancer cells in a dose- and time-dependent manner. The upregulation of EVs production in cancer cells after exposure to ionizing radiation are related to the p53 signaling pathway which activated by DNA damage. These findings suggest that radiation can enhance the release of EVs from cancer cells, which may aid in developing cancer treatment strategies [7, 9]. Similarly, nonionizing radiation (e.g. ultraviolet radiation) [10], photodynamic therapy (using Foscan® photosensitizer) [11], and acoustic irradiation (using low power and high frequency) [12] have been reported to boost the number of EVs secreted by cancer cells. Furthermore, low-level electrical stimulation applied to murine melanoma and fibroblast cells has been found to stimulate EV secretion, possibly by activating Rho guanosine triphosphatase (GTPase). In another study,

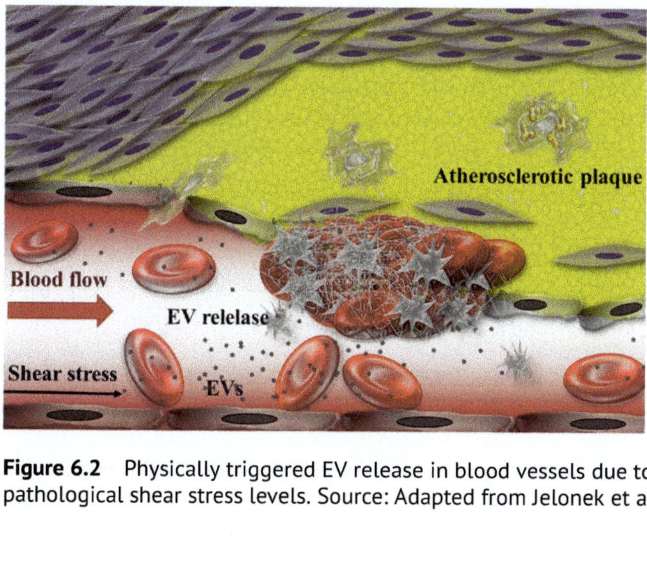

Figure 6.2 Physically triggered EV release in blood vessels due to physiological and pathological shear stress levels. Source: Adapted from Jelonek et al. [7].

focal and transient electrical stimulation-induced nano-perforation of the cell membrane was shown to increase EV secretion from mouse embryonic fibroblasts by 50 times [13]. The use of magnetic iron(III) oxide nanoparticles (Fe_3O_4) and a static magnetic field (SMF) combination was also shown to enhance the small EV secretion of BM-MSCs. The Fe_3O_4+SMF group yielded the highest amount of EVs, followed by the Fe_3O_4 group and the untreated group. Functionally, the small EVs secreted by the Fe_3O_4+SMF group showed greater osteogenesis and angiogenesis potential than those secreted by the Fe_3O_4 group or naive cells [14]. Moreover, the incorporation of nanoparticles has been tested in various studies to increase the amount of EVs secreted by cultured cells. For instance, platinum nanoparticles were found to increase the secretion of EVs in human lung epithelial adenocarcinoma cells by inducing oxidative stress and ceramide [15]. Similarly, silver–titanium oxide nanoparticles could increase the secretion of EVs in B16-F10 mouse melanoma cells by inducing oxidative stress [16], while calcium phosphate particles could promote the secretion of EVs in macrophage-like RAW264.7 cells and monocyte-like THP-1 cells by promoting the formation of multivesicular bodies and their fusion with plasma [17].

6.1.3 Chemical Stimulation

Numerous studies have reported the addition of chemicals to the culture medium to promote the secretion of EVs. For example, cancer cells and renal explants have been found to increase the number of small EVs secreted when treated with sodium iodoacetate (IAA), a glycolysis inhibitor, and 2,4-dinitrophenol (DNP), an oxidative phosphorylation inhibitor. In vivo experiments involving mice injected with IAA/DNP verified these findings, demonstrating increased small EVs in the blood. Wang et al. investigated the effects of several small molecules on the production of small EVs in brain-derived mesenchymal stem cells [18]. They discovered that fenoterol, norepinephrine, *N*-methyldopamine, mefenasen, and forskolin stimulate the secretion of small EVs, with the concentration of each molecule affecting the degree of increase. Norepinephrine combined with forskolin and norepinephrine combined with *N*-methyldopamine showed synergistic effects on exocrine secretion, whereas *N*-methyldopamine combined with forskolin did not. Additionally, Suxiao Jiuxin Pill, a traditional Chinese herbal medicine, demonstrated a synergistic effect in promoting the secretion of small EVs from cardiac mesenchymal stem cells via the GTPase-dependent pathway [19]. Adiponectin, a fat factor, increased the number of small EVs produced by MSCs and vascular endothelial cells through T-cadherin [20], as well as the concentration of EVs in the serum of mice, which enhanced the cardiac protection function of mesenchymal stem cells. Finally, studies suggest that intracellular calcium ions regulate the release of EVs, and exposure to calcium has been shown to increase the production of EVs [17].

6.1.4 Physiological Modification

The secretion and function of EVs by cells may vary depending on the microenvironment. Under physiological conditions, the oxygen levels in peripheral tissues, known as "hypoxia," typically range between 1% and 11% [21]. Studies have suggested that hypoxic preconditioning can enhance the therapeutic effects of mesenchymal stem cells [22] by increasing the production of EVs. For instance, research has shown that UC-MSCs cultured under hypoxia (5% O_2) proliferate and remain more viable than those cultured under normal oxygen (21% O_2). Hypoxic UC-MSCs produce more small EVs that have been proven more effective at alleviating chronic airway inflammation and lung remodeling in asthmatic mice induced by ovalbumin [22]. Additionally, the small EVs secreted by hypoxic BM-MSCs have been found to have more therapeutic potential in promoting cartilage and spinal cord regeneration than those produced by normal hypoxic cells [23]. Furthermore, the angiogenesis potential of EVs secreted by endothelial cells cultured under hypoxia is better than that of cells cultured under normoxic conditions [24]. Extremely hypoxic conditions (0.5% O_2) also lead to an increase in the release of small EVs from MSCs [25]. In terms of function, hypoxic small EVs have been shown to promote myocardial repair more effectively than normoxic small EVs, as they can promote vascularization, reduce myocardial cell apoptosis, reduce scar tissue formation, and enhance the recruitment of cardiac progenitor cells. Thermal stimulation has also been found to enhance EV production. For example, Hedlund et al. [26] and Clayton et al. [27] reported an increase in EV secretion from cells when cultured at 40 °C for 3 hours. Harmati et al. found that B16-F10 mouse melanoma cells produce more small EVs when cultured at 42 °C for 3 times for 2 hours [16], and Gong et al. found that MGC-803 human gastric cancer cells released more small EVs at high temperatures (40 °C) [28]. pH values are crucial in many biological processes and cell metabolism [29]. Kim et al. reported that even slight differences in the pH of the culture medium can affect cell reprogramming and differentiation [30]. Apart from hypoxia, extracellular acidity is another sign of cancer due to the accumulation of glycolytic metabolites such as lactic acid [31]. Melanoma cells, for example, secrete more Lamp-2$^+$, CD81$^+$, and Rab5B$^+$ EVs under acidic conditions (pH = 6) to transport tumor-related proteins to other cells [32]. Gong et al. also reported that MGC-803 human gastric cancer cells secrete more small EVs at acidic pH (pH = 4) [28]. Overall, cells can adapt to environmental stress and pathological conditions such as those found in the tumor microenvironment (i.e. high temperature, low oxygen tension, and low pH level) by modifying the production of EVs. The changes in EVs production are also being studied to gain a better understanding of how cells respond to stress and pathological conditions.

6.1.5 Genetic Manipulation

The signal pathway of EV secretion in parental cells was genetically modified. Rab protein, a GTPase that regulates vesicular flow, was found to regulate the secretion pathway of EVs [33]. Bobrie et al. [34] and Ostrowski et al. [35] demonstrated that silencing the RAB27A and RAB27B genes reduces the docking between vesicular endosome (MVE) and plasma membrane, resulting in reduced small EV secretion in cancer cells [36, 37]. The recent studies have also revealed that Rab13 and Rab7a regulate EV secretion in cancer cells [38, 39]. Phospholipase D (PLD) catalyzes the hydrolysis of phosphatidylcholine to produce phosphatidic acid, which is an important lipid messenger that participates in cell signal transduction, including extracellular and endocytosis. Overexpression of PLD2 in RBL-2H3 cells increased the secretion of EVs labeled by BODYPI ceramide, while knockout of PLD2 decreased EV release. PLD2 was found to be an effector of the ADP ribosylation factor 6 (*ARF*6) gene, regulating the sprouting of intracavitary vesicles (ILV) and playing an indispensable role in the production of EVs [40]. Moreover, *ARF*6 can control the shedding of microbubbles in tumor cells [41]. Lentivirus transduction of cysteamine CD9 into HEK293, SH-SY5Y, HeLa, Raji, and Jurkat cell lines increased small EV secretion up to three times [42]. P2X7 and SNAREs receptor's expression were also found to affect the formation, transport, and secretion of EVs in cells [43, 44]. Stress was shown to regulate EV production by cells [45]. Cell immortalization technology was used to achieve consistent production of cell sources needed for large-scale use of EVs. Highly amplifiable human ESC-MSCs were produced by transfection of the MYC gene, bypassing cell senescence and maintaining high proliferation [46]. Small EVs secreted by immortalized MYC-transformed hESC-MSCs showed potential for heart protection in the model of myocardial ischemia–reperfusion injury *in vivo* [47]. Daily injection of small EVs secreted by immortalized MYC-MSC was found to be safe and did not affect tumor growth. However, the influence of EVs on cancer progression deserves further study, because different studies have reported different results [48].

6.2 Extraction of EVs

EVs are a vital form of membrane-bound vesicle (ranging from 30 to 150 nm in diameter) that can be secreted by almost any cell. They serve as an essential medium for intercellular communication and epigenetic regulation, playing significant roles under both pathological and physiological conditions. EVs carry finely regulated cargo from parent cells, conveying specific information to other cells, while simultaneously allowing us to infer the function and status of their source cells based on the type and quantity of their contents. EVs are readily accessible *in vitro*, making liquid biopsy using EVs a potential for noninvasive disease diagnosis. Furthermore, EVs offer unique advantages in treating diseases, including drug delivery, development of cell-free vaccines, and regenerative medicine. Consequently, the potential of EVs has attracted substantial research and capital investments. To further the research and development of EVs, it is necessary to advance related technologies such as production, isolation, and quality control. Although there is no standard production technology for EVs, researchers have developed several methods and techniques to isolate them by exploring their biochemical and physicochemical characteristics.

There are two main categories of EVs based on their origins: animal-derived EVs and plant-derived EVs. The former can be obtained from diverse cell cultures such as fibroblasts, immune cells, adipocytes, mesenchymal stem cells, and tumor cells. Additionally, they can be isolated from a variety of body fluids, including blood, urine, saliva, breast milk, and amniotic fluid. Moreover, animal-derived EVs can be directly extracted from tissues such as the spleen, brain, and muscle.

6.2.1 Separation Strategies of EVs

Extensive research has been conducted on the potential of EVs as a diagnostic and therapeutic tool for disease. However, the progress of this field relies heavily on technological advancements. There are currently two major technical obstacles that hinder the development of EVs: (i) the need for simplified extraction techniques and increased production yields; and (ii) efficient identification of EVs. Despite significant efforts, there is still no standardized method for isolating EVs. Researchers have explored various methods based on the biochemical and physical characteristics of EVs. For example, ultracentrifugation is considered the "gold standard" due to its high processing capacity. However, EV samples prepared using this method can be contaminated with protein aggregates and lipoproteins, leading to inaccuracies in quantification and functional analysis. As of now, there is no universally applicable method for isolating EVs. To address this issue, researchers have developed six categories of EV isolation strategies based on the physicochemical and biochemical properties of different EVs. These categories include ultracentrifugation, ultrafiltration, immunoaffinity capture, polymer precipitation based on charge neutralization, size-exclusion chromatography (SEC), and microfluidics. Each technique has its own unique advantages and disadvantages.

6.2.2 Ultracentrifugation Approach

Ultracentrifugation is widely regarded as the "gold standard" for extracting EVs and is the most commonly used method. As early as 1987, Johnston employed ultracentrifugation to isolate EVs from the culture medium of reticulocyte tissue [49]. The technique involves separating various extracellular components in a fluid sample based on their density, size, and shape under a specific centrifugal force. Unlike other methods, ultracentrifugation does not require additional separation reagents, significantly reducing the risk of contamination from the original sample and lowering costs. This makes it an ideal technique for large-scale EV production. However, the method does have strict requirements for the centrifuge equipment, which must be highly accurate and speedy. Each centrifugation process can take more than ten hours or longer, necessitating continuous operation by preparatory personnel. Additionally, high-speed centrifugation can cause mechanical damage and protein aggregation to EVs. Thus, ultracentrifugation is not suitable for clinical diagnostic applications with small sample sizes due to its equipment and inherent characteristics.

6.2.2.1 Differential Ultracentrifugation

The precipitation method, also known as simple ultracentrifugation, is the most commonly used strategy for isolating EVs. The process of purifying EVs via ultracentrifugation is illustrated in Figure 6.3. The first step involves the continuous centrifugation of the sample at increasing speeds to eliminate large dead cells and cell debris.

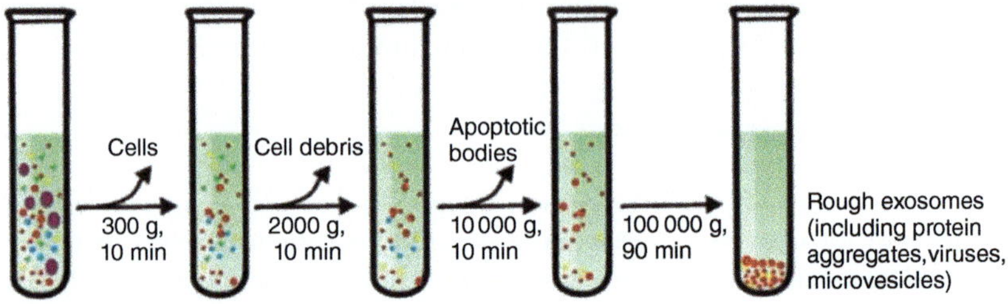

Figure 6.3 Schematic diagram of exosome separation based on differential ultracentrifugation. Source: Yang et al. [50]/Ivyspring International Publisher/CC BY 4.0.

Discard the precipitate and use the supernatant for the subsequent step. Eventually, ultracentrifuge the supernatant at 100 000 × g, and the pellet that forms corresponds to the EVs. Wash the precipitate with a large volume of PBS to eliminate any contaminating proteins, and perform a final high-speed centrifugation. According to He et al., ultracentrifugation was utilized to isolate EVs derived from ovarian cancer cells, and it was discovered that ovarian cancer EVs were transferred from cancer cells to endothelial cells via the miR-205 signaling molecule [51]. In brief, differential ultracentrifugation necessitates many centrifugation cycles to sequentially eliminate impurities like cell debris, apoptotic bodies, and protein aggregates.

Differential ultracentrifugation has become a popular method for isolating EVs from various sources, such as cell culture media, serum, saliva, urine, and cerebrospinal fluid, due to its ease of use, lack of specialized technical knowledge required, and ability to prepare large volumes without complex sample pretreatment [52]. However, the use of differential centrifugation for EV isolation often results in poor purity due to the similarity in sedimentation properties of different types of EVs. To address this issue, Muller et al. have proposed a refined approach that involves using ultracentrifugation to isolate EVs from plasma, followed by ultrafiltration on the differential centrifugation plasma/serum to remove plasma contaminants, finally, performing SEC to achieve high-purity samples. While differential ultracentrifugation remains a reliable and efficient method for large-scale preparation of EVs, the proposed method by Muller et al. can significantly increase sample purity [53].

The application of the same centrifugation protocol to different rotors without considering differences in solution viscosity often leads to dissimilar data obtained by different research groups. This heterogeneity may arise due to differences in proportions of EVs and shedding vesicles. To address this issue, Livshts et al. performed a theoretical analysis of the sedimentation of vesicles using differential centrifugation techniques with different rotor types. Their theory considers various parameters, such as geometric parameters, speed, centrifugation duration, and solution viscosity of the rotor, allowing for the evaluation of the proportion of particle-like vesicles with specific sizes and densities. By conducting relatively simple theoretical calculations, a "stand-alone" differential ultracentrifugation protocol suitable for the rotor used can be selected [54].

6.2.2.2 Isopycnic and Moving-Zone Density Gradient Ultracentrifugation

Density gradient ultracentrifugation is a widely used method for isolating extracellular components, such as EVs, apoptotic bodies, and protein aggregates. The technique involves layering biocompatible media with varying densities, such as iodixanol or sucrose, in a test tube, with the density gradually decreasing from the bottom to the top. The sample is then added to the top of the density gradient medium, and the mixture is centrifuged for an extended period (e.g. 100 000 × g, for 16 hours). This process allows for the separation of components with different buoyant densities. EVs, for instance, remain in the culture medium layer, while protein aggregates concentrate at the bottom of the centrifuge tube. Through sucrose density gradient centrifugation, purified EV samples can be obtained for downstream applications. Researchers, such as Zhu et al., have used density gradient centrifugation to isolate and characterize nicotine-treated macrophage-derived EVs, including detecting mRNA levels [55]. With a popularity of around 11.6% among currently used EV isolation methods, density gradient ultracentrifugation has proven to be a reliable and effective method for obtaining pure EV samples.

Density ultracentrifugation is a widely used technique for isolating EVs, but its specificity is not very high; so different density gradient ranges must be utilized depending on the sample [56]. Moving boundary ultracentrifugation, on the other hand, has the unique characteristic of using a culture medium with lower density than all solutes in the sample, resulting in the separation of all solutes not only based on density but also based on mass/size. This allows for the separation of vesicles with comparable densities but different sizes, such as EVs, viruses, and large microvesicles [57]. Pérez-González et al. successfully isolated and identified EVs in brain tissue using density ultracentrifugation [58]. However, careful determination of the centrifugation time is required to minimize the precipitation of extracellular particles. To this end, high-density culture medium is usually added as a buffer at the bottom of the centrifuge tube.

Ultracentrifugation is a highly effective method for obtaining samples with exceptional purity. Unfortunately, the need for specialized equipment and trained personnel has impeded its widespread use and further development. Moreover, the mechanical forces generated by high-speed centrifugation can have irreversible effects on the structure and biological functions of EVs, thereby hindering their modification and subsequent use in drug development [59]. Thus, it is imperative to explore alternative methods of EVs isolation that preserve their biological functions.

6.2.3 Ultrafiltration Approach

Ultrafiltration is a filtration method that employs ultrafine nanomembranes with different molecular weight cutoffs (MWCO) to separate EVs from clinical samples or cell culture media. This method distinguishes EVs from co-vesicles based on size, similar to conventional filtration techniques. Numerous easy-to-use ultrafiltration devices have been developed that can achieve rapid EVs preparation and yield

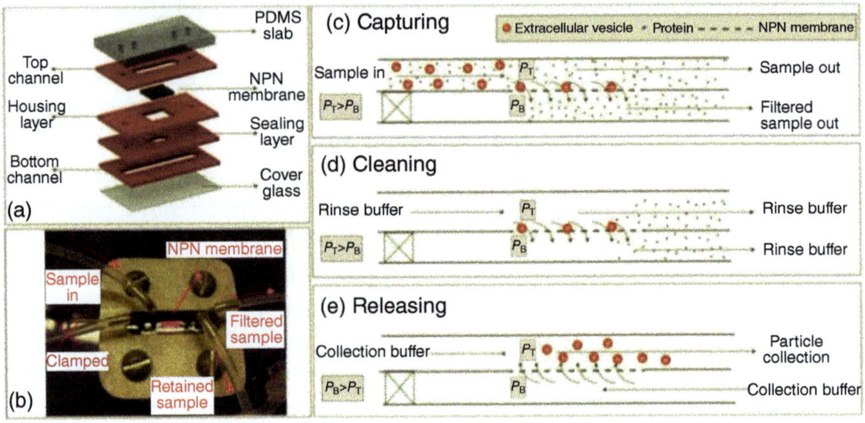

Figure 6.4 Tangential Flow Analyte capture (TFAC) technique for particle separation. Source: Lebreton et al. [61]/with permission of John Wiley & Sons, Inc.

comparable to ultracentrifugation. For instance, Dehghani et al. developed a series-connected microfilter (see Figure 6.4) that comprises two series-connected microfilters with defined size exclusion limits of approximately 20–200 nm. The pores of the ultrafiltration membrane capture EVs and similar-sized analytes, while the EVs are washed in the membrane and subsequently released through additional flow [60].

Ultrafiltration of sample liquids can lead to clogging of the membrane due to high viscosity, reducing the lifespan of the membrane. To overcome this issue, protease treatment can be employed, although tangential flow filtration (TFF) technology is considered a superior alternative. TFF involves the parallel flow of the sample liquid and membrane, allowing partial flow to pass through the membrane under controlled fluid dynamics. This method reduces the risk of clogging, as the membrane is continuously flushed and potential blockages are minimized. The remaining material can be recycled back to the feed container for repeated filtration, leading to an automated program and high yield [61]. TFF-based EVs preparation has already been used in clinical trials [62], with promising results seen in the recent studies where dendritic cell-derived EVs prepared using TFF effectively promoted T-cell responses for anticancer therapy [63].

Ultrafiltration has gained popularity due to its efficiency and ease of operation, accounting for approximately 5.4% of currently used EVs isolation methods. Unlike ultracentrifugation, which takes 16 hours, ultrafiltration takes only a few minutes and does not require special equipment. This method offers numerous advantages, including low equipment cost, fast operation, and portability. While EVs obtained using this method have medium purity, nanometer-sized particles that are comparable in size to EVs are not excluded. However, the method has its challenges, including shear stress caused by nanomembranes and centrifugal force that can lead to deformation and rupture of EVs, clogging of nanomembranes, high costs, and low service life of nanomembranes. Additionally, retention of nanomembranes may result in yield loss and lower recovery rates.

6.2.4 Size-Exclusion Chromatography

Size-exclusion chromatography (SEC) is a valuable analytical technique that separates solutes by comparing the pore size of the gel with the size of the sample molecules. The mobile phase elutes large molecules that cannot enter the gel pores, whereas small molecules are retained in the column and eluted more slowly (Figure 6.5) [50]. The biggest advantage of SEC is that it preserves the natural biological activity of the separated EVs. Unlike ultracentrifugation and ultrafiltration, SEC operates through passive gravity flow, which does not damage vesicle structure or integrity. Selecting elution buffers with physiological osmolarity and viscosity can further maintain the natural state of EVs. In addition to preserving the biological function of EVs, SEC offers other advantages. It requires a small sample volume, and commercially available SEC columns can process volumes as small as 15 μl to achieve high resolution, standardized, and reproducible EV separation, making it ideal for fingertip EV analysis. EV collection using SEC is simple, compatible with various types of fluids, and usually requires no additional pretreatment steps. The entire process is time efficient and labor saving, and selective porous materials and buffer systems enable the generation of a defined subpopulation of EVs. Compared with ultrafiltration, SEC has minimal sample loss and high recovery rates. This method is not only applicable to small liquid samples but can also be scaled up and automated for high-throughput EV preparation. Shu et al. compared several traditional EV isolation methods and analyzed the cytokine spectrum by multiplex bead array, demonstrating that the number of soluble factors and all the constituents of EVs prepared depended on the separation method used. The combination of ultrafiltration and SEC produced 58 times than ultracentrifugation [64]. However, one disadvantage of the SEC method is that it produces EVs with a wide size distribution, especially in the low-size range, and is unable to exclude pollutants with similar sizes to EVs.

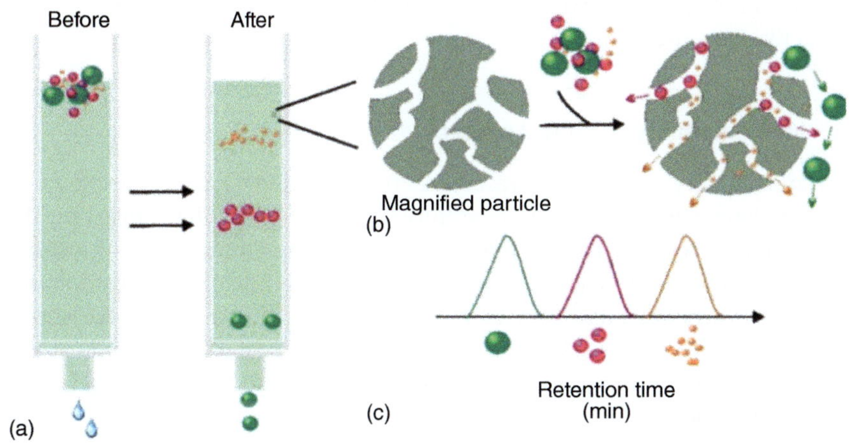

Figure 6.5 EVs separation principle based on size exclusion chromatography. Source: Yang et al. [50]/Ivyspring International Publisher/CC BY 4.0.

6.2.5 Polymer Precipitation Strategy

The process of polymer precipitation relies on the interaction between incredibly hydrophilic polymers and the water molecules encompassing EVs. This interaction leads to the creation of a hydrophobic microenvironment, prompting the precipitation of EVs. To begin the process, pretreatment is necessary to eliminate large contaminant particles such as apoptotic bodies and cell debris. The pretreated sample is then incubated with a PEG solution overnight at 4 °C. Finally, the EVs are precipitated and collected via low-speed centrifugation (1500 × g).

Gall et al. have introduced an improved polymer precipitation method for extracting a significant amount of EVs without requiring as much differentiation into myocytes as ultracentrifugation. This approach entails a modified polymer-based precipitation technique, accompanied by additional washing steps, to optimize EV yield from lower numbers of differentiated myocytes and less conditioned medium. By avoiding senescence, this approach allows for multiple experiments to be performed without depleting myocyte proliferation capacity. To execute this method, clarified medium is mixed with Total EVs Isolation reagent (LifeTechnologies™) at a 2 : 1 volume ratio and left to incubate overnight at 4 °C. Afterward, the mixture is centrifuged at 10 000 g for 1 hour at 4 °C, and the resulting pellet is resuspended in 500 μl of PBS. The EVs are washed thrice using a 100 kDa Amicon® filter column and finally resuspended in 100 μl of PBS or NuPAGE™ LDS sample buffer for protein blotting experiments [65].

Polymer precipitation offers a simple and effective method to upscale production to large quantities, yielding impressive results with minimal equipment requirements. Additionally, this approach presents a promising avenue for swift disease diagnosis via the integration of diverse EVs detection platforms. However, certain concerns arise when employing PEG for EVs precipitation. These include inadequate rates of purity and recovery, elevated levels of nonspecific proteins leading to false positives, nonuniform distribution of particle sizes, generation of stubborn polymers, and the possibility of mechanical or chemical disruption of EVs through additives like Tween-20.

6.2.6 Immunoaffinity Capture Approach

EVs invariably carry proteins and receptors. To isolate EVs from specific sources, researchers have devised an immunoaffinity capture method that hinges on the selective binding between labeled foreign substances and immobilized antibodies (ligands). This approach yields high-purity EVs that are free of chemical impurities, effortless to handle, and easy to use. Nonetheless, this method necessitates expensive antibodies, necessitates the optimization of EVs labeling, and may entail low processing volume and yield. What's more, extra steps to elute EVs may impair their inherent structure.

Submicron magnetic particles are widely used in immunoprecipitation methods that require recombinant proteins. This method offers high sensitivity and captures efficiency owing to its large surface area and uniformity. It is highly adaptable, and allows for amplification or reduction of specific applications, making it ideal for processing large sample volumes. EVs are characterized by specific surface markers such as CD63 and CD9 proteins that can be isolated and separated by incubating them with magnetic beads coated with anti-labeled antibodies. In a novel approach, Wang et al. extracted and eluted foreign substances by utilizing magnetic separation and photoactivated cargo release with spatiotemporal control. In this system, magnetic beads were comodified with photosensitive groups: nitrobenzyl and aptamers that were compatible with CD63—a highly expressed surface-specific protein of foreign substances [66]. Morphological and proteomic analysis of EVs in cell models and nude mouse tumor transplantation models demonstrated that the new magnetic bead system outperformed the current ultracentrifugation method in terms of extraction time, yield, and proportion of high-expression CD63 populations.

The magnetic bead method is a convenient and specific technique that preserves EVs' integrity. However, its efficiency is low, and EVs can be affected by pH and salt concentration, which complicates downstream experiments and hinders its widespread adoption. Yang et al. have developed a new, highly effective approach to capture EVs by utilizing the affinity interaction between PS and CLIKKPF—a high-affinity peptide. The team immobilized CLIKKPF on a silicon dioxide microsphere (SiO_2-pep) surface and used this material to capture foreign substance samples from Hela cell cultures. They compared the efficiency of this method with three classical methods that use serum samples (UF, UC, DGC). The SiO_2-pep affinity method has great potential for expansion and value, as it was successfully used to isolate EVs from healthy individuals, hepatocellular carcinoma (HCC), and intrahepatic bile duct cancer (CCA) patient's sera. By comparing the proteomic maps of EVs, accurate and minimally invasive biomarkers for early and differential diagnosis of HCC and CCA can be identified, and potential molecular mechanisms of these two liver cancers can be further explored [67].

6.2.7 Microfluidic Technology

Immunomicrofluidic devices for EVs separation have the same principle as immunoaffinity separation methods. The method involves recognizing labeled foreign substances through antibodies fixed on a chip, providing high efficiency, low cost, portability, ease of automation, and integrated diagnosis. However, due to the fixed chips, the sample volume that can be accommodated is limited.

Microfluidic methods for EVs separation can be classified into two categories: label free and affinity based. Label-free methods use physical properties, such as size, density, rigidity, and electrical properties, to isolate EVs from nontargeted biological particles within the microfluidic device. This can be achieved through passive (filtration, inertial/viscoelastic separation, deterministic lateral displacement, pinched flow separation, and flow field-flow separation) or active (electric, acoustic, and magnetic separation) means, driven by EVs' separation forces. These label-free microfluidic methods enable isolation of high-resolution and high-yield EVs from various biological fluids at low cost and with simplicity.

On the other hand, affinity-based microfluidic separation employs probes modified with micro-/nanostructures to capture EVs with specific surface proteins or lipids within the microfluidic device [68].

With the advent of precision machining technology, researchers have been able to explore contactless charging as a mechanism for particle sorting, including elastic lift, acoustic, and bidirectional electrophoresis, in addition to antibody- and scale-dependent microfluidics. These methods allow for the efficient, scalable, and high-quality isolation of EVs. However, most of the existing EV separation technologies are mainly applied in the basic research. To make microfluidic technology more clinically relevant, a more translational approach is required in designing techniques and equipment for EV separation. This involves thoroughly evaluating a sufficient number of clinical samples to improve selectivity, robustness, and sensitivity.

One such technique is the size-dependent microfluidic chip developed by Han et al. for tangential flow filtration-based separation and purification of EVs. The microfluidic chip is composed of two symmetrical layers of polymethyl methacrylate (PMMA) with serpentine channels and a nanoporous polycarbonate track-etched (PCTE) membrane located between them. The PMMA layer has a stable internal structure to enhance the stability of the microchip (Figure 6.6) [69]. The PCTE membrane contains uniformly cylindrical and precise pores to ensure efficient removal of protein contaminants.

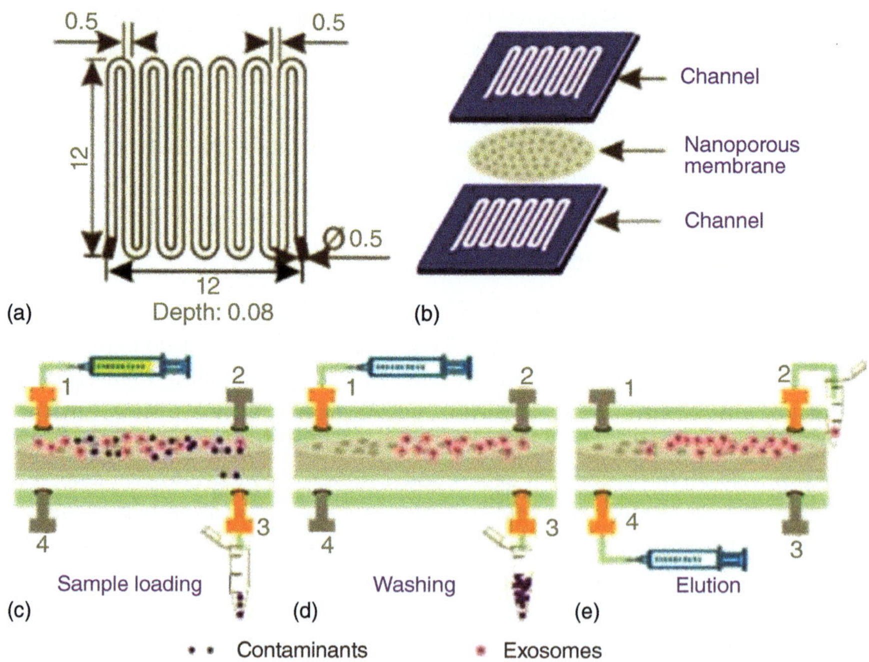

Figure 6.6 Schematic diagram of EVs microfluidic separation method. Source: Han et al. [69]/with permission of Elsevier.

This method was first validated in a mixture of HeLa cell-derived EVs and BSA, and subsequently validated in real human plasma samples. Under optimized microfluidic parameters, the removal rate of contaminants was greater than 97%, and the recovery rate of EVs was greater than 80%. Compared to UC, this method has the advantages of low cost, simplicity of operation, low sample volume, fast processing speed, high recovery rate of EVs, and high efficiency of contaminant removal. Additionally, tangential flow filtration can largely avoid microfluidic clogging compared to other filtration-based foreign substance purification methods.

6.2.8 Other Methods

Pan et al. introduced EV-FISHER—a MOF-based platform for rapid enrichment and separation of EVs. The platform comprises a UiO-66-NH_2 MOF with defects and a cleavable lipid probe—PSDC, which is a hybrid of DNA and cholesterol spaced by PO_4^{3-}. In Figure 6.7 the authors showed that the defect UiO-66-NH_2 was activated

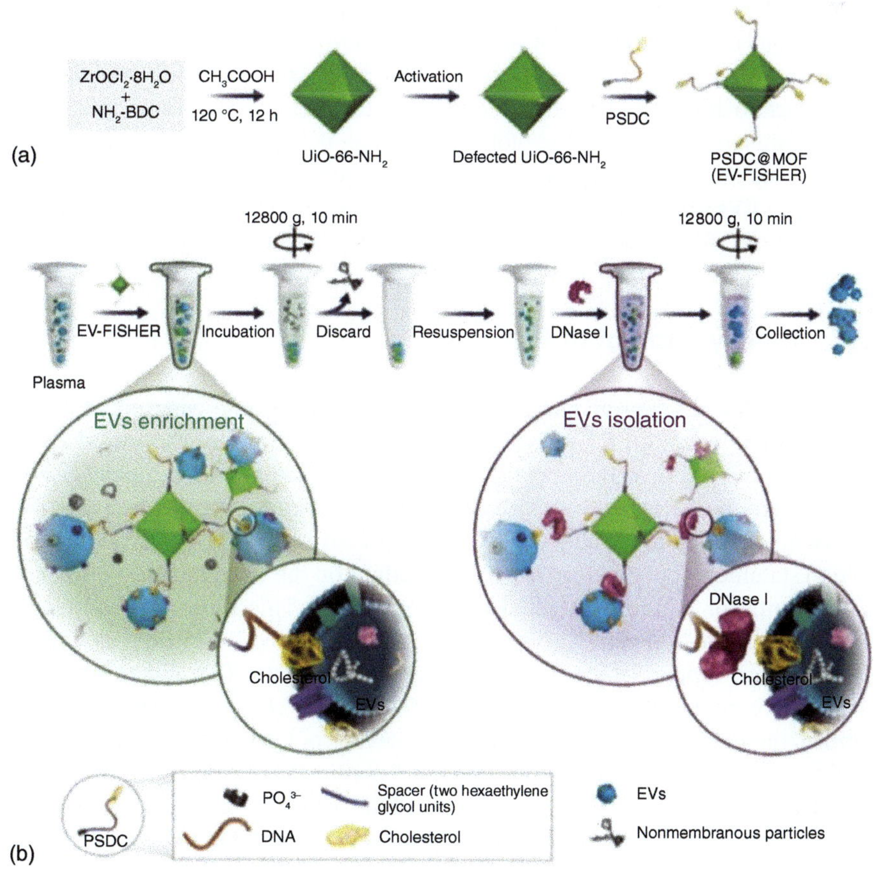

Figure 6.7 Schematic diagram of (a) EV-FISHER synthesis and (b) plasma EVs separation workflow. Source: Pan et al. [70]/John Wiley & Sons.

to increase the modification site, then combined with PSDC through PO_4^{3-}–Zr(IV) to form EV-FISHER (PSDC@MOF) with a simple and high-affinity connection. The MOF acts as a "handle" to bind with the "fishing line" PSDC, which is baited with cholesterol to capture EVs in plasma. The captured EVs were enriched by low-speed centrifugation, and then DNase I was added to hydrolyze the DNA in PSDC, effectively separating the EVs. Finally, the purified EVs were obtained after removing the MOF fragments. Compared to the traditional ultracentrifugation method, EV-FISHER exhibited superior performance in terms of time (40 minutes versus 240 minutes), separation efficiency (74.2% versus 18.1%), and separation requirement (12800 g versus 135 000 g), owing to the high affinity of cholesterol for EV membranes and the advantages of MOF. The authors demonstrated that EV-FISHER is a promising tool for EV isolation and enrichment [70].

In spite of the significant advancements in EVs research achieved over the last few decades, the development of effective strategies for isolating EVs continues to be a major unresolved issue. This is largely attributable to the intricate nature of biological fluids, with EVs, lipoproteins, viruses, and other EVs displaying considerable overlap in physicochemical and biochemical properties, in addition to the heterogeneity of EVs themselves. Consequently, there is presently no single isolation technique for EVs that is universally applicable to all studies. Even the gold standard ultracentrifugation method is frequently hampered by protein and lipoprotein contaminants, depending on the biological sample being used. In such scenarios, the judicious employment of two or more techniques provides a rational strategy for effective EVs isolation, such as the synergistic use of immunoaffinity-based EVs capture (or ultrafiltration) and density gradient centrifugation.

6.3 Quality Control of EVs

The use of EVs in scientific and clinical research has become widespread due to their immense potential for early disease diagnosis and prognosis, drug delivery systems, and innovative therapeutic formulations. However, as a relatively new biological product, there are currently no specific regulations or guidance principles in place for ensuring the quality control of EVs. In 2014, the ISEV board published a position statement that provided expert suggestions on the "minimum experimental requirements for defining EVs and their functions." This statement included a minimum information list for EVs research, which covered EVs' isolation, characterization, and functional studies. As EVs are considered cell products, guidance principles for quality control can be derived from those related to cell products. For instance, the FDA released the "Content and Review of Chemistry, Manufacturing, and Control (CMC) Information for Human Somatic Cell Therapy Investigational New Drug Applications (INDs)" in 2016, which provides detailed guidance on biologic safety and product quality stability for various materials added during the production process, such as cell banks. Additionally, the ICH's Q5A(R1), Q5A(R2) EWG provides guidance on detecting exogenous viruses in cell products, while the National Medical Products Administration issued the "Technical Guidelines for

Research and Evaluation of Cell Therapy Products" and related interpretations in 2017, which guide the evaluation of risks associated with the use of serums and other materials, cell banks, and other evaluations for cell therapy and corresponding products.

Based on this, many researchers are committed to developing scientific methods for specific quality control of EVs [71].

6.3.1 Transmission Electron Microscopy

Transmission Electron Microscopy (TEM) is a widely adopted technique for analyzing EVs, replacing traditional optical microscopes due to their smaller diameter than the optical wavelength [48]. TEM uses short-wavelength electrons to capture the characteristics and morphology of EVs (Figure 6.8), and its application has become increasingly widespread. However, due to the complexity of sample pretreatment and preparation operations required for TEM analysis, this method is not suitable

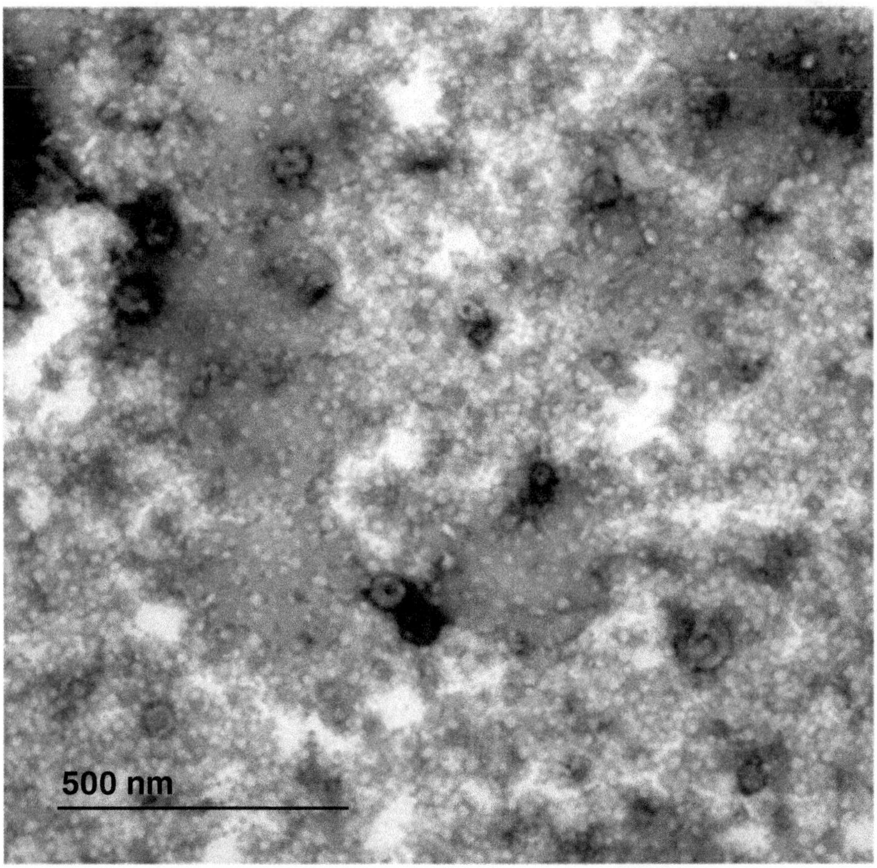

Figure 6.8 TEM image of cup-shaped EVs with a diameter smaller than 100 nm. Source: Zhang et al. [72]/with permission of John Wiley & Sons, Inc.

for rapid measurement of large quantities of EVs, which may result in inaccurate measurement of EVs concentration. Researchers typically use other techniques, such as nanoparticle tracking analysis (NTA), to detect the size or quantity of EVs and supplement these results with TEM to visualize the quality of EVs [72].

6.3.2 High-Resolution Liquid Chromatography–Mass Spectrometry

High-resolution liquid chromatography–mass spectrometry (LC–MS) is a powerful tool used for lipid and protein analysis of EVs. Given the mechanism of EV formation, the distribution of lipids on the membrane of EVs is expected to be linked to the composition of the plasma membrane, including phospholipids, sphingolipids, and cholesterol. In a study by Tsai et al., the expression profiles of EV proteins from the aqueous humor of nine highly myopic patients and nine control patients were compared using liquid chromatography–tandem mass spectrometry (LC–MS-MS). Apolipoprotein A1 (APOA1) and optocin (OPTC) were among the proteins analyzed to assess changes in EV protein in myopic patients [73]. Although LC–MS offers high sensitivity and a broad detection range, it requires skilled personnel and expensive instruments for metabolomics research. Furthermore, careful sample preparation and complex data processing and analysis steps limit its application in routine quality control for EVs [74].

6.3.3 Enzyme-Linked Immunosorbent Assay

The Enzyme-linked immunosorbent assay (ELISA) is a valuable tool for determining protein adsorption and content in EVs. This rapid, sensitive, and highly repeatable detection method is widely used for detecting and quantifying EVs. Logozzi et al. developed a sandwich ELISA utilizing tumor-related markers CD63, Rab-5b, and Caveolin-1 to detect and quantify EVs in cell culture media and plasma samples (Figure 6.9) [75]. Additionally, researchers have created ExoELISA kits (System Biosciences) based on direct ELISA, where EV particles and proteins

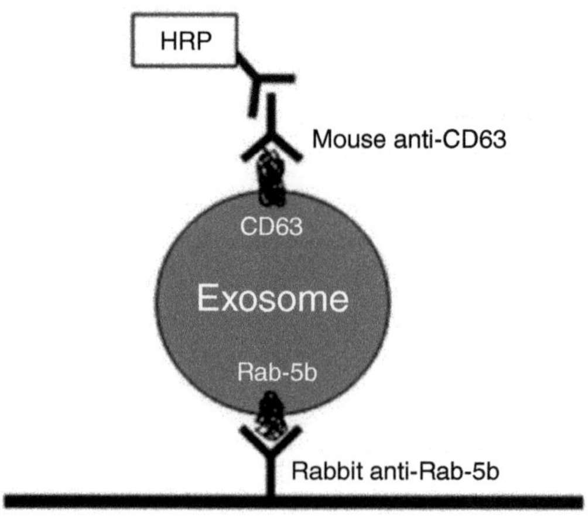

Figure 6.9 Schematic diagram of the Exotest–an ELISA used for EVs detection and quantification. Source: Logozzi et al. [75]/PLOS/CC BY 4.0.

are directly immobilized in microwells, and blocked after binding to prevent nonspecific binding of detection antibodies. The detection antibodies are then added to the sample wells and bind to specific antigen proteins on EVs (such as CD63, CD9, and CD81). Multiple tests are often necessary due to differences in protein expression in various body fluids and cell lines, resulting in a large sample size requirement and variations between different batches of testing.

6.3.4 Fourier-Transform Infrared Attenuated Total Reflection Spectroscopy

Fourier-transform infrared attenuated total reflection spectroscopy (ATR-FTIR) is a highly effective method that utilizes functional groups and spectral bands to provide both qualitative and quantitative descriptions of individual EVs. This technique allows for rapid detection of protein and lipid content on EVs, providing crucial biochemical information, and can be utilized for very small sample volumes (2 μl) to obtain relative changes in EV composition. Therefore, ATR-FTIR can be considered the standard colorimetric analysis for EV quality analysis. Although less specific and sensitive than MS, it provides valuable spectral information that allows for quick assessment of the reproducibility or repeatability between isolated EVs, and provides information on differences in their lipid content. In the recent study, researchers utilized ATR-FTIR to analyze EVs extracted from human milk, separated by ultracentrifugation after single-phase extraction, and conducted lipidomics analysis using UPLC-QqTOF-MS/MS and automated annotation based on various databases. Multivariate analysis showed significant correlations between ATR-FTIR-specific regions and concentrations of different lipid classes, and ATR-FTIR provided a qualitative description of EV lipid content [73]. Additionally, ATR-FTIR spectroscopy has been used to measure the protein concentration of red blood cell-derived EVs (REVs) and calculate the integrated area of the amide I band based on the IR spectrum of REVs. This measurement was found to be proportional to the protein content in the sample and unrelated to its secondary structure. Spike-in and dilution experiments were carried out to determine the ability of ATR-FTIR spectroscopy to quantify protein in EV samples, and the results showed that ATR-FTIR technology provides a reliable method for non-reagent-based protein quantification of EVs (see Figure 6.10) [76].

6.3.5 Capillary Electrophoresis

Capillary electrophoresis (CE) is a liquid-phase separation technique that utilizes capillaries as separation channels and high-voltage electric fields as driving forces to separate various components in a sample based on their distribution behavior. The use of CE in EV quality control provides quantitative information about EVs and can be combined with dark-field microscopy to detect individual vesicles in microfluidic channels to estimate ζ potential [77]. Kato et al. demonstrated that microcapillary electrophoresis on chips and laser dark-field microscopy can accurately evaluate the ζ potential distribution of individual EVs. CE is therefore

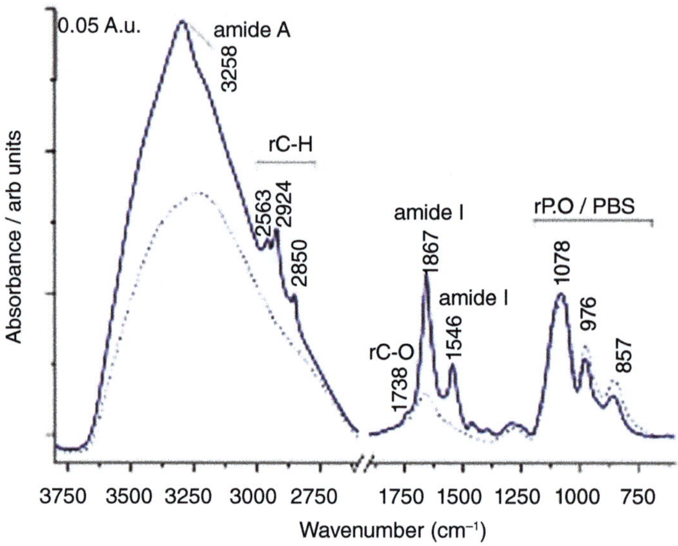

Figure 6.10 ATR spectrum obtained from diluting samples of REVs. Source: Szentirmai et al. [76]/Springer Nature/CC BY 4.0.

well-suited for tracking the electrophoretic migration of single EVs [78]. Akagi et al. equipped a micro-capillary electrophoresis system on a chip with a laser dark-field microscope and discovered that cancer cell-derived EVs had a larger ζ potential than normal cells. After neuraminidase treatment, the ζ potential of EVs derived from normal prostate cells and prostate cancer cells were evaluated using the system, revealing that cancer cell-derived EVs had abundant negative charge due to large amounts of sialic acid. These findings suggest that the micro-capillary electrophoresis system on a chip is a reliable method to evaluate surface events occurring on single EVs [79].

6.3.6 Nanoparticles Tracking Analysis

Nanoparticle tracking analysis (NTA) is a remarkable technology that allows real-time visualization and direct analysis of nanoparticles present in liquids. This technique is particularly useful for rapidly evaluating the size and concentration of EVs (see Figure 6.11). During NTA measurements, a focused laser beam illuminates the particles, and the light scattered by each particle is focused by a microscope onto an image sensor. The camera's charge coupling captures the scattered light from moving particles, including nanovesicles such as EV particles. A special type of NTA software uses the Stokes–Einstein equation to relate motion to particle size [80]. This technique has been used by researchers such as Andreu et al. [81] and Helwa et al. [82] to evaluate the purity and quantity of EVs' samples for quality control purposes, and to compare the efficiency of different schemes and commercial kits for isolating EVs.

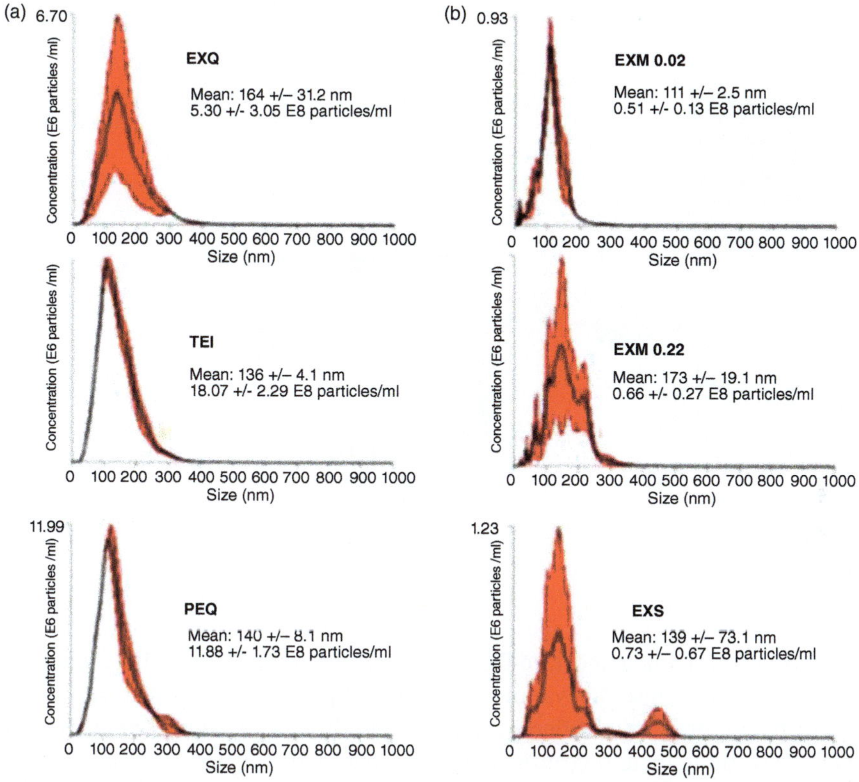

Figure 6.11 NTA measured the mean value, concentration, and average particle size of EV samples. Source: Adapted from Tian et al. [84].

In a groundbreaking experiment, Pasalic et al. used the fluorescence properties of NTA to measure the amount of miR-21 target contained in lung cancer-derived EVs and its chemical stoichiometry relative to the total EVs population. This involved using a mixture of tumor cell-derived EVs and cationic lipid complex nanoparticles fused with fluorescently labeled small RNA (miRNA)-specific molecular beacons. These probes were fluorescence-quenched oligonucleotide hybridization probes whose fluorescence was restored upon binding to the target nucleic acid sequence. After mixing EVs with particles carrying the molecular beacon, a combination of light scattering and fluorescence nanoparticle tracking was used to identify the proportion of the total EVs' population containing the target miRNA transcript. This groundbreaking experiment opens up new avenues for identifying specific targets carried by vesicles and demonstrates the exciting potential of NTA technology in biomedical research [83].

6.3.7 Flow Cytometer

The diameter of EVs, on average, is about 100 nm. However, a flow cytometer's laser beam cannot distinguish the light scattered by particles smaller than 300 nm.

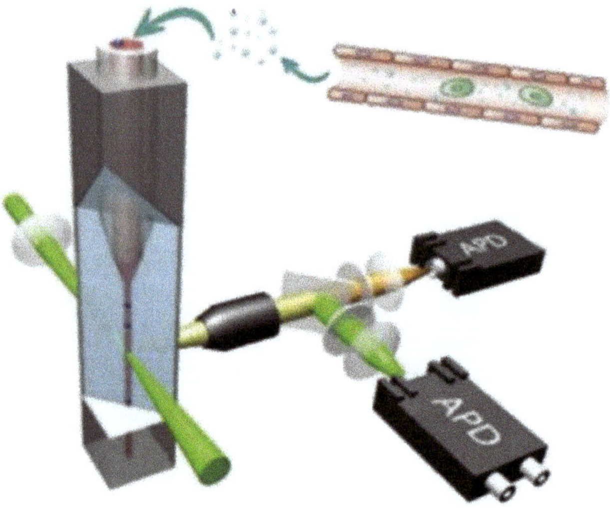

Figure 6.12 Schematic diagram of the nFCM constructed in the laboratory. Source: Tian et al. [84]/Taylor & Francis.

This necessitates fixing nanosized EVs onto microbeads when analyzing them using a flow cytometer. Presently, various microbead types, such as those composed of latex or polystyrene, with different sizes ranging from 4 to 9 μm in diameter, and functionalizations, including antibodies, streptavidin–biotin proteins, formaldehyde sulfate, have been designed for this purpose. These studies show that coupling EVs to microbeads allows visualization using flow cytometry, enabling specific surface protein analysis, overcoming size limitations. However, this method's practical use is limited, as a large amount of starting material or preenriched EVs is required before experimentation. Researchers are actively exploring new methods to establish a standardized and reproducible approach for detecting EVs by flow cytometry. For instance, Tian et al. used a laboratory-manufactured nanoscale flow cytometer (nFCM) capable of analyzing single EVs as small as 40 nm, establishing a new benchmark for evaluating the quality of EVs isolated from plasma [84] (Figure 6.12). Meanwhile, Liu et al. reported a pH-mediated assembly system that aggregates single nanosized EVs into micrometer-sized clusters directly analyzed using conventional flow cytometry, overcoming the size limit of EVs analysis [85].

6.3.8 Other Techniques

Researchers have recently explored several new nanotechnologies for detecting or separating EVs, such as resistive pulse sensing (RPS), surface plasmon resonance (SPR) nanosensors, and Nano-DLD (deterministic lateral displacement). Prior to the development of Nano-DLD, the smallest biological particle that could be diagnosed by size was circulating tumor cell, which is roughly 50 times larger than EVs. Nano-DLD offers uniform gap sizes ranging from 25 to 235 nm, and it is produced using a silicon manufacturing process by a collaboration of scientists from IBM, Mount Sinai Icahn School of Medicine, and Princeton University. Nano-DLD allows for the detection of EVs in the 20–100 nm size range, which was not achievable using the previous methods. The recent literature suggests that cryo-electron microscopy (cryo-EM) combined with receptor-specific gold labeling can be utilized for the phenotypic analysis of selected EV subpopulations, which can determine the physical and chemical properties of EVs, including their morphology, size, and features of the lipid bilayer. For instance, Arraud et al. employed cryo-EM and receptor-specific gold standards to distinguish different types of EVs in plasma, along with each EV-specific source, highlighting the diversity of EVs in pure plasma [86].

The quality of EVs is intricately linked to the quality of their host cells, as they are a byproduct of cellular metabolism. Hence, developing a reproducible, large-scale, and high-throughput GMP-level method for their preparation is of utmost importance for the advancement and utilization of EV-based therapies. Numerous studies have focused on optimizing GMP-level production processes for EVs, establishing quality monitoring protocols, and creating standardized techniques for clinical-scale EVs preparation. Andriolo et al. have designed a GMP-level method for large-scale preparation of Exo-CPCs—a drug derived from heart progenitor cells (CPCs), with the goal of evaluating the safety, identity, and potency of the final product [87]. To ensure the quality of EVs, GMP requirements are primarily centered around three aspects: upstream cell culture systems, downstream purification systems, and quality control of the final EVs product.

The upstream cell culture system must create the appropriate environment for secretion and excretion [88], while downstream purification of EVs involves three key steps: filtration to remove cell debris, concentration of the culture medium, and isolation of EVs from the concentrated medium. Although differential centrifugation is a common strategy for concentrating EVs [89], it demands a lot of labor, which has led to the increased popularity of tangential flow filtration (TFF)—a less time-consuming alternative. Studies have found that EVs purified using TFF have higher immune regulatory efficacy and contain more soluble factors [90]. SEC is another method developed for purification of EVs, which focuses on recovery rate and specificity [91]. While differential centrifugation achieves high purity of EVs, each step of the process reduces the recovery rate, and it is time-consuming. Sucrose gradient centrifugation overcomes the long processing time of differential centrifugation while maintaining high purity, but residual sucrose reagents remain a concern. Ultrafiltration systems offer improved specificity and recovery rates, while also overcoming time consumption issues, but can cause instability in proteins.

Currently, there is an enthusiastic effort underway in both domestic and foreign research regarding EVs. Despite the existing mysteries surrounding EVs, and the challenges faced in developing and applying methods for quality control, EVs have attracted increasing attention in disease diagnosis, microenvironment regulation mechanisms, and drug carrier transformation. As the exploration of EVs deepens, and research tools and methods continue to evolve, a more standardized system for the extraction, isolation, and quality control of EVs will be established. By resolving issues related to EVs extraction, isolation, and quality control, the potential of EVs being used as therapeutic drugs in clinical settings will greatly increase.

References

1 Dionysios, C.W., Defne, B., Avinash, S. et al. (2016). Efficient production and enhanced tumor delivery of engineered extracellular vesicles. *Biomaterials* 105: 195–205.

2 Guo, S., Debbi, L., Zohar, B. et al. (2021). Stimulating extracellular vesicles production from engineered tissues by mechanical forces. *Nano Lett.* 21 (6): 2497–2504.

3 Yan, L. and Wu, X. (2019). Exosomes produced from 3D cultures of umbilical cord mesenchymal stem cells in a hollow-fiber bioreactor show improved osteochondral regeneration activity. *Cell Biol. Toxicol.* 36: 165–178.

4 Cao, J., Wang, B., Tang, T. et al. (2020). Three-dimensional culture of MSCs produces exosomes with improved yield and enhanced therapeutic efficacy for cisplatin-induced acute kidney injury. *Stem Cell Res. Ther.* 11: 206.

5 Gobin, J., Muradia, G., Mehic, J. et al. (2021). Hollow-fiber bioreactor production of extracellular vesicles from human bone marrow mesenchymal stromal cells yields nanovesicles that mirrors the immuno-modulatory antigenic signature of the producer cell. *Stem Cell Res. Ther.* 12: 127.

6 Kim, M., Yun, H.-W., Park, D.Y. et al. (2018). Three-dimensional spheroid culture increases exosome secretion from mesenchymal stem cells. *Tissue Eng. Regener. Med.* 15: 427–436.

7 Jelonek, K., Widlak, P., and Pietrowska, M. (2016). The influence of ionizing radiation on exosome composition, secretion and intercellular communication. *Protein Pept. Lett.* 23 (7): 656–663.

8 Piffoux, M., Nicolás-Boluda, A., Mulens-Arias, V. et al. (2018). Extracellular vesicles for personalized medicine: the input of physically triggered production, loading and theranostic properties. *Adv. Drug Delivery Rev.* 138: 247–258.

9 Pavlakis, E., Neumann, M., and Stiewe, T. (2020). Extracellular vesicles: messengers of p53 in tumor–stroma communication and cancer metastasis. *Int. J. Mol. Sci.* 21 (24): 9648.

10 Arscott, W.T., Tandle, A.T., Zhao, S. et al. (2014). Ionizing radiation and glioblastoma exosomes: implications in tumor biology and cell migration. *Transl. Oncol.* 6 (6): 638–648.

11 Aubertin, K., Silva, A.K.A., Luciani, N. et al. (2016). Massive release of extracellular vesicles from cancer cells after photodynamic treatment or chemotherapy. *Sci. Rep.* 6: 35376.

12 Ambattu, L.A., Ramesan, S., Dekiwadia, C. et al. (2020). High frequency acoustic cell stimulation promotes exosome generation regulated by a calcium-dependent mechanism. *Commun. Biol.* .

13 Fukuta, T., Nishikawa, A., and Kogure, K. (2020). Low level electricity increases the secretion of extracellular vesicles from cultured cells. *Biochemis. Biophys. Rep.* 21: 100713.

14 Wu, D., Chang, X., Tian, J. et al. (2021). Bone mesenchymal stem cells stimulation by magnetic nanoparticles and a static magnetic field: release of exosomal miR-1260a improves osteogenesis and angiogenesis. *J. Nanobiotechnol.* 19: 209.

15 Gurunathan, S., Kang, M.-H., Jeyaraj, M. et al. (2021). Platinum nanoparticles enhance exosome release in human lung epithelial adenocarcinoma cancer cells (A549): oxidative stress and the ceramide pathway are key players. *Int. J. Nanomed.* 16: 515–538.

16 Harmati, M., Gyukity-Sebestyen, E., Dobra, G. et al. (2019). Small extracellular vesicles convey the stress-induced adaptive responses of melanoma cells. *Sci. Rep.* 9: 15329.

17 Shyong, Y.-J., Chang, K.-C., and Lin, F.-H. (2018). Calcium phosphate particles stimulate exosome secretion from phagocytes for the enhancement of drug delivery. *Colloids Surf., B* 171: 391–397.

18 Wang, J., Bonacquisti, E.E., Brown, A.D. et al. (2020). Boosting the biogenesis and secretion of mesenchymal stem cell-derived exosomes. *Cells* 9 (3): 660.

19 Ruan, X-f., Ju C-w., Shen, Y. et al. (2018). Suxiao Jiuxin pill promotes exosome secretion from mouse cardiac mesenchymal stem cells in vitro. *Acta Pharmacol. Sin.* 39: 569–578.

20 Obata, Y., Kita, S., Koyama, Y. et al. (2018). Adiponectin/T-cadherin system enhances exosome biogenesis and decreases cellular ceramides by exosomal release. *JCI Insight* 3 (8): e99680.

21 Carreau, A., El Hafny-Rahbi, B., Matejuk, A. et al. (2011). Why is the partial oxygen pressure of human tissues a crucial parameter? Small molecules and hypoxia. *J. Cell. Mol. Med.* 15 (6): 1239–1253.

22 Kim, Y., Jin, H.J., Heo, J. et al. (2018). Small hypoxia-primed mesenchymal stem cells attenuate graft-versus-host disease. *Leukemia* 32: 2672–2684.

23 Rong, Y., Zhang, J., Jiang, D. et al. (2020). Hypoxic pretreatment of small extracellular vesicles mediates cartilage repair in osteoarthritis by delivering miR-216a-5p. *Acta Biomater.* 122: 325–342.

24 Zhang, H.-C., Liu, X.-B., Huang, S. et al. (2012). Microvesicles derived from human umbilical cord mesenchymal stem cells stimulated by hypoxia promote angiogenesis both in vitro and in vivo. *Stem Cells Dev.* 21 (18): 3289–3297.

25 Zhu, J., Lu, K., Zhang, N. et al. (2017). Myocardial reparative functions of exosomes from mesenchymal stem cells are enhanced by hypoxia treatment of the cells via transferring microRNA-210 in an nSMase2-dependent way. *Artif. Cells Nanomed. Biotechnol.* 46 (8): 1659–1670.

26 Hedlund, M., Nagaeva, O., Kargl, D. et al. (2011). Thermal- and oxidative stress causes enhanced release of NKG2D ligand-bearing immunosuppressive exosomes in leukemia/lymphoma T and B cells. *PLoS One* 6 (2): e16899.

27 Clayton, A., Turkes, A., Navabi, H. et al. (2005). Induction of heat shock proteins in B-cell exosomes. *J. Cell Sci.* 118 (16): 3631–3638.

28 Gong, C., Zhang, X., Shi, M. et al. (2021). Tumor exosomes reprogrammed by low pH are efficient targeting vehicles for smart drug delivery and personalized therapy against their homologous tumor. *Adv. Sci.* 8 (10): 2002787.

29 Aoi, W. and Marunaka, Y. (2014). Importance of pH homeostasis in metabolic health and diseases: crucial role of membrane proton transport. *Biomed. Res. Int.* 4: 598986.

30 Kim, N., Minami, N., Yamada, M. et al. (2017). Immobilized pH in culture reveals an optimal condition for somatic cell reprogramming and differentiation of pluripotent stem cells. *Reprod. Med. Biol.* 16 (1): 58–66.

31 Swietach, P., Vaughan-Jones, R.D., Harris, A.L. et al. (2014). The chemistry, physiology and pathology of pH in cancer. *Philos. Trans. R. Soc. B: Biol. Sci.* 369 (1638): 20130099.

32 Parolini, I., Federici, C., Raggi, C. et al. (2009). Microenvironmental pH is a key factor for exosome traffic in tumor cells. *J. Biol. Chem.* 284 (49): 34211–34222.

33 Goody, R.S., Müller, M.P., and Wu, Y.-W. (2017). Mechanisms of action of Rab proteins, key regulators of intracellular vesicular transport. *Biol. Chem.* 398 (5-6): 565–575.

34 Bobrie, A., Krumeich, S., Reyal, F. et al. (2012). Rab27a supports exosome-dependent and -independent mechanisms that modify the tumor microenvironment and can promote tumor progression. *Cancer Res.* 72 (19): 4920–4930.

35 Ostrowski, M., Carmo, N.B., Krumeich, S. et al. (2010). Rab27a and Rab27b control different steps of the exosome secretion pathway. *Nat. Cell Biol.* 12: 19–30.

36 Guerra, F., Paiano, A., Migoni, D. et al. (2019). Modulation of RAB7A protein expression determines resistance to cisplatin through late endocytic pathway impairment and extracellular vesicular secretion. *Cancers* 11 (1): 52.

37 Thakur, R., Panda, A., Coessens, E. et al. (2016). Phospholipase D activity couples plasma membrane endocytosis with retromer dependent recycling. *eLife*.

38 Thakur, R., Naik, A., Panda, A. et al. (2019). Regulation of membrane turnover by phosphatidic acid: cellular functions and disease implications. *Front. Cell Dev. Biol.* 7: 83.

39 Laulagnier, K., Grand, D., Dujardin, A. et al. (2004). PLD2 is enriched on exosomes and its activity is correlated to the release of exosomes. *FEBS Lett.* 572 (1-3): 11–14.

40 Muralidharan-Chari, V., Clancy, J., Plou, C. et al. (2009). ARF6-regulated shedding of tumor cell-derived plasma membrane microvesicles. *Curr. Biol.* 19 (22): 1875–1885.

41 Böker, K.O., Lemus-Diaz, N., Rinaldi Ferreira, R. et al. (2017). The impact of the CD9 tetraspanin on lentivirus infectivity and exosome secretion. *Mol. Ther.* 26 (2): 634–647.

42 Qu, Y. and Dubyak, G.R. (2009). P2X7 receptors regulate multiple types of membrane trafficking responses and non-classical secretion pathways. *Purinergic Signalling* 5: 163–173.

43 Hessvik, N.P. and Llorente, A. (2017). Current knowledge on exosome biogenesis and release. *Cell. Mol. Life Sci.* 75: 193–208.

44 Kucharzewska, P. and Belting, M. (2013). Emerging roles of extracellular vesicles in the adaptive response of tumour cells to microenvironmental stress. *J. Extracell. Vesicles* 2 (1): 20304.

45 Chen, T.S., Arslan, F., Yin, Y. et al. (2011). Enabling a robust scalable manufacturing process for therapeutic exosomes through oncogenic immortalization of human ESC-derived MSCs. *J. Transl. Med.* 9: 47.

46 Tan, T.T., Lai, R.C., Padmanabhan, J. et al. (2021). Assessment of tumorigenic potential in mesenchymal-stem/stromal-cell-derived small extracellular vesicles (MSC-sEV). *Pharmaceuticals* 14 (4): 345.

47 Rodini, C.O., Gonçalves da Silva, P.B., Assoni, A.F. et al. (2018). Mesenchymal stem cells enhance tumorigenic properties of human glioblastoma through independent cell-cell communication mechanisms. *Oncotarget* 9 (37): 24766–24777.

48 Arraud, N., Linares, R., Tan, S. et al. (2014). Extracellular vesicles from blood plasma: determination of their morphology, size, phenotype and concentration. *J. Thromb. Haemost.* 74 (5): 1844–1851.

49 Johnstone, R.M., Bianchini, A., and Teng, K. (1989). Reticulocyte maturation and exosome release: transferrin receptor containing exosomes shows multiple plasma membrane functions. *Blood* 9 (26): 8206–8220.

50 Yang, D., Zhang, W., Zhang, H. et al. (2020). Progress, opportunity, and perspective on exosome isolation - efforts for efficient exosome-based theranostics. *Theranostics* 10 (8): 3684–3707.

51 He, L., Zhu, W., Chen, Q. et al. (2019). Ovarian cancer cell-secreted exosomal miR-205 promotes metastasis by inducing angiogenesis. *Theranostics* 9 (26): 8206–8220.

52 Tran, P.H.L., Wang, T., Yin, W. et al. (2019). Development of a nanoamorphous exosomal delivery system as an effective biological platform for improved encapsulation of hydrophobic drugs. *Int. J. Pharm.* 566: 697–707.

53 Muller, L., Hong, C.-S., Stolz, D.B. et al. (2014). Isolation of biologically-active exosomes from human plasma. *J. Immunol. Methods* 411: 55–65.

54 Livshits, M.A., Khomyakova, E., Evtushenko, E.G. et al. (2015). Isolation of exosomes by differential centrifugation: theoretical analysis of a commonly used protocol. *Sci. Rep.* 5: 17319.

55 Zhu, J., Liu, B., Wang, Z. et al. (2019). Exosomes from nicotine-stimulated macrophages accelerate atherosclerosis through miR-21-3p/PTEN-mediated VSMC migration and proliferation. *Theranostics* 9 (23): 6901–6919.

56 Paolini, L., Zendrini, A., Noto, G.D. et al. (2016). Residual matrix from different separation techniques impacts exosome biological activity. *Sci. Rep.* 6: 23550.

57 Chen, B.-Y., Sung, C.W.-H., Chen, C. et al. (2019). Advances in exosomes technology. *Clin. Chim. Acta*.

58 Crescitelli, R., Lasser, C., and Lotvall, J. (2021). Isolation and characterization of extracellular vesicle subpopulations from tissues. *Nat. Protoc.* 16 (3): 1548–1580.

59 Van Deun, J., Mestdagh, P., Sormunen, R. et al. (2014). The impact of disparate isolation methods for extracellular vesicles on downstream RNA profiling. *J. Extracell. Vesicles* 3.

60 Dehghani, M., Lucas, K., Flax, J. et al. (2019). Tangential flow microfluidics for the capture and release of nanoparticles and extracellular vesicles on conventional and ultrathin membranes. *Adv. Mater. Technol.* 4 (11): 1900539.

61 Lebreton, B., Brown, A., and van Reis, R. (2008). Application of high-performance tangential flow filtration (HPTFF) to the purification of a human pharmaceutical antibody fragment expressed in *Escherichia coli*. *Biotechnol. Bioeng.* 100: 964–974.

62 Escudier, B., Dorval, T., Chaput, N. et al. (2005). Vaccination of metastatic melanoma patients with autologous dendritic cell (DC) derived-exosomes: results of thefirst phase I clinical trial. *J. Transl. Med.* 3(10).

63 Besse, B., Charrier, M., Lapierre, V. et al. (2016). Dendritic cell-derived exosomes as maintenance immunotherapy after first line chemotherapy in NSCLC. *OncoImmunology* 5 (4): e1071008.

64 Shu, S.L., Yang, Y., Allen, C.L. et al. (2019). Purity and yield of melanoma exosomes are dependent on isolation method. *J. Extracell. Vesicles* 10 (1): 20.

65 Le Gall, L., Ouandaogo, Z.G., Anakor, E. et al. (2020). Optimized method for extraction of exosomes from human primary muscle cells. *Skelet Muscle* 10 (1): 20.

66 Wang, C., Zhang, D., Yang, H. et al. (2022). A light-activated magnetic bead strategy utilized in spatio-temporal controllable exosomes isolation. *Front. Bioeng. Biotechnol.* 10: 1006374.

67 Yang, K., Jia, M., Cheddah, S. et al. (2022). Peptide ligand-SiO(2) microspheres with specific affinity for phosphatidylserine as a new strategy to isolate exosomes and application in proteomics to differentiate hepatic cancer. *Bioact. Mater.* 15: 343–354.

68 Vaidyanathan, R., Naghibosadat, M., Rauf, S. et al. (2014). Detecting exosomes specifically: a multiplexed device based on alternating current electrohydrodynamic induced nanoshearing. *Anal. Chem.* 86 (22): 11125–11132.

69 Han, Z., Peng, C., Yi, J. et al. (2021). Highly efficient exosome purification from human plasma by tangential flow filtration based microfluidic chip. *Sens. Actuators, B* 333: 129563.

70 Pan, W.L., Feng, J.J., Luo, T.T. et al. (2022). Rapid and efficient isolation platform for plasma extracellular vesicles: EV-FISHER. *J. Extracell. Vesicles* 11 (11): e12281.

71 Lötvall, J., Hill, A.F., Hochberg, F. et al. (2014). Minimal experimental requirements for definition of extracellular vesicles and their functions: a position statement from the International Society for Extracellular Vesicles. *J. Extracell. Vesicles* 3: 26913.

72 Zhang, M., Jin, K., Gao, L. et al. (2018). Methods and technologies for exosome isolation and characterization. *Small Methods* 2 (9): 1800021.

73 Tsai, C.-Y., Chen, C.-T., Lin, C.-H. et al. (2021). Proteomic analysis of exosomes derived from the aqueous humor of myopia patients. *Int. J. Med. Sci.* 18 (9): 2023.

74 Victoria, R.-G., Isabel, T.-D., Alba, M.-G. et al. (2021). ATR-FTIR spectroscopy for the routine quality control of exosome isolations. *Chemom. Intell. Lab. Syst.* 217: 104401.

75 Logozzi, M., De Milito, A., Lugini, L. et al. (2009). High levels of exosomes expressing CD63 and caveolin-1 in plasma of melanoma patients. *PLoS One* 4 (4): e5219.

76 Szentirmai, V., Wacha, A., Németh, C. et al. (2020). Reagent-free total protein quantification of intact extracellular vesicles by attenuated total reflection Fourier transform infrared (ATR-FTIR) spectroscopy. *Anal. Bioanal.Chem.* 412: 4619–4628.

77 Steć, A., Jońca, J., Waleron, K. et al. (2022). Quality control of bacterial extracellular vesicles with total protein content assay, nanoparticles tracking analysis, and capillary electrophoresis. *Int. J. Mol. Sci.* 23 (8): 4347.

78 Kei, K., Masashi, K., Nami, H. et al. (2013). Electrokinetic evaluation of individual exosomes by on-chip microcapillary electrophoresis with laser dark-field microscopy. *Jpn. J. Appl. Phys.* 52 (6S): 06GK10.

79 Akagi, T., Kato, K., Hanamura, N. et al. (2014). Evaluation of desialylation effect on zeta potential of extracellular vesicles secreted from human prostate cancer cells by on-chip microcapillary electrophoresis. *Jpn. J. Appl. Phys.* 53 (6S): 06JL1.

80 Boriachek, K., Islam, M.N., Möller, A. et al. (2017). Biological functions and current advances in isolation and detection strategies for exosome nanovesicles. *Small* 14 (6): 1702153.

81 Andreu, Z., Rivas, E., Sanguino-Pascual, A. et al. (2016). Comparative analysis of EV isolation procedures for miRNAs detection in serum samples. *J. Extracell. Vesicles* 5 (1): 31655.

82 Helwa, I., Cai, J., Drewry, M.D. et al. (2017). A comparative study of serum exosome isolation using differential ultracentrifugation and three commercial reagents. *PLoS One* 12 (1): e0170628.

83 De Sousa, K.P., Rossi, I., Abdullahi, M. et al. (2022). Isolation and characterization of extracellular vesicles and future directions in diagnosis and therapy. *WIREs Nanomed. Nanobiotechnol.* 15 (1): e1835.

84 Tian, Y., Gong, M., Hu, Y. et al. (2019). Quality and efficiency assessment of six extracellular vesicle isolation methods by nano-flow cytometry. *J. Extracell. Vesicles* 9 (1): 1697028.

85 Liu, X., Zong, Z., Xing, M. et al. (2021). pH-Mediated clustering of exosomes: breaking through the size limit of exosome analysis in conventional flow cytometry. *Nano Lett.* 21 (20): 8817–8823.

86 Linares, R., Tan, S., Gounou, C. et al. (2015). High-speed centrifugation induces aggregation of extracellular vesicles. *J. Extracell. Vesicles* 4 (1): 29509.

87 Andriolo, G., Provasi, E., Lo Cicero, V. et al. (2018). Exosomes from human cardiac progenitor cells for therapeutic applications: development of a gmp-grade manufacturing method. *Front. Physiol.* 9: 1169.

88 Harn, H.-J., Chen, Y.-S., Lin, E.-Y. et al. (2020). Exosomes in clinical trial and their production in compliance with good manufacturing practice. *Tzu Chi Med. J.* 32 (2): 113.

89 Gupta, S., Rawat, S., Arora, V. et al. (2018). An improvised one-step sucrose cushion ultracentrifugation method for exosome isolation from culture supernatants of mesenchymal stem cells. *Stem Cell Res. Ther.* 9: 1–11.

90 Bari, E., Perteghella, S., Catenacci, L. et al. (2019). Freeze-dried and GMP-compliant pharmaceuticals containing exosomes for acellular mesenchymal stromal cell immunomodulant therapy. *Nanomedicine* 14 (6): 753–765.

91 Watson, D.C., Yung, B.C., Bergamaschi, C. et al. (2018). Scalable, cGMP-compatible purification of extracellular vesicles carrying bioactive human heterodimeric IL-15/lactadherin complexes. *J. Extracell. Vesicles* 7 (1): 1442088.

7

Prospects and Limitations of Clinical Application of Extracellular Vesicles

Li Luo[1,2,3], *Weirun Li*[1,2], *and Zhenhua Li*[1,2,3]

[1] *The Tenth Affiliated Hospital of Southern Medical University, 78 Wanjiang Avenue, Dongguan, Guangdong 523059, China*
[2] *Southern Medical University, The First School of Clinical Medicine, Shatai South Avenue, Guangzhou, Guangdong 510080, China*
[3] *Guangdong Provincial Key Laboratory of Cardiac Function and Microcirculation, Shatai South Avenue, Guangzhou, Guangdong 510080, China*

7.1 Application of Exosomes as Liquid Biopsy in Clinical Diagnosis

Exosomes are a key type of EV, which were originally described to be released from sheep reticulocytes [1]. With the large number of studies that followed, exosomes were found to exist in almost all body fluids, primarily blood [2], urine [3], cerebrospinal fluid [4], saliva [5], pleural effusion [6], ascites fluid [7], amniotic fluid [8], breast milk [9], and bronchoalveolar lavage fluid (BALF) [10]. Exosomes, originating from the endosomal pathway via the formation of late endosomes or multivesicular bodies, enclose a variable spectrum of molecules characterized by parent cells, including nucleic acids [DNA, mRNA, microRNA (miRNA), lncRNA, circRNA, etc.], proteins, and lipids, which can be transported over distances within the protection of a lipid bilayer-enclosed structure [11]. Exosomes have a great promise to serve as novel biomarkers in liquid biopsy, since large quantities of exosomes are enriched in body fluids, and are involved in numerous physiological and pathological processes [12].

7.2 Exosomes—It has Become a Star Molecule in Disease Diagnosis

Exosomes are vesicles secreted by cells with particle sizes ranging from 30 to 200 nm and carrying a variety of biological active molecules such as proteins, nucleic acids, lipids, etc. They have the following characteristics: By delivering the biologically molecules carried to the receptor cells, the exosomes can mediate the information exchange between cells; the contents of exosomes are cell- specific and vary according to the status of the source cells, which can reflect the physiological or

Biomedical Applications of Extracellular Vesicles, First Edition.
Edited by Zhenhua Li, Xing-Jie Liang, and Ke Cheng.

pathological status of the source cells. Of the most interest is that the exosomes are found in all biological fluids and are secreted by all cells, rendering them attractive as minimally invasive liquid biopsies with the potential for longitudinal sampling to follow disease progression. Based on the above biological characteristics, exosome biogenesis enables the capture of a complex extracellular and intracellular molecular cargo for comprehensive, multiparameter diagnostic testing (Figure 7.1).

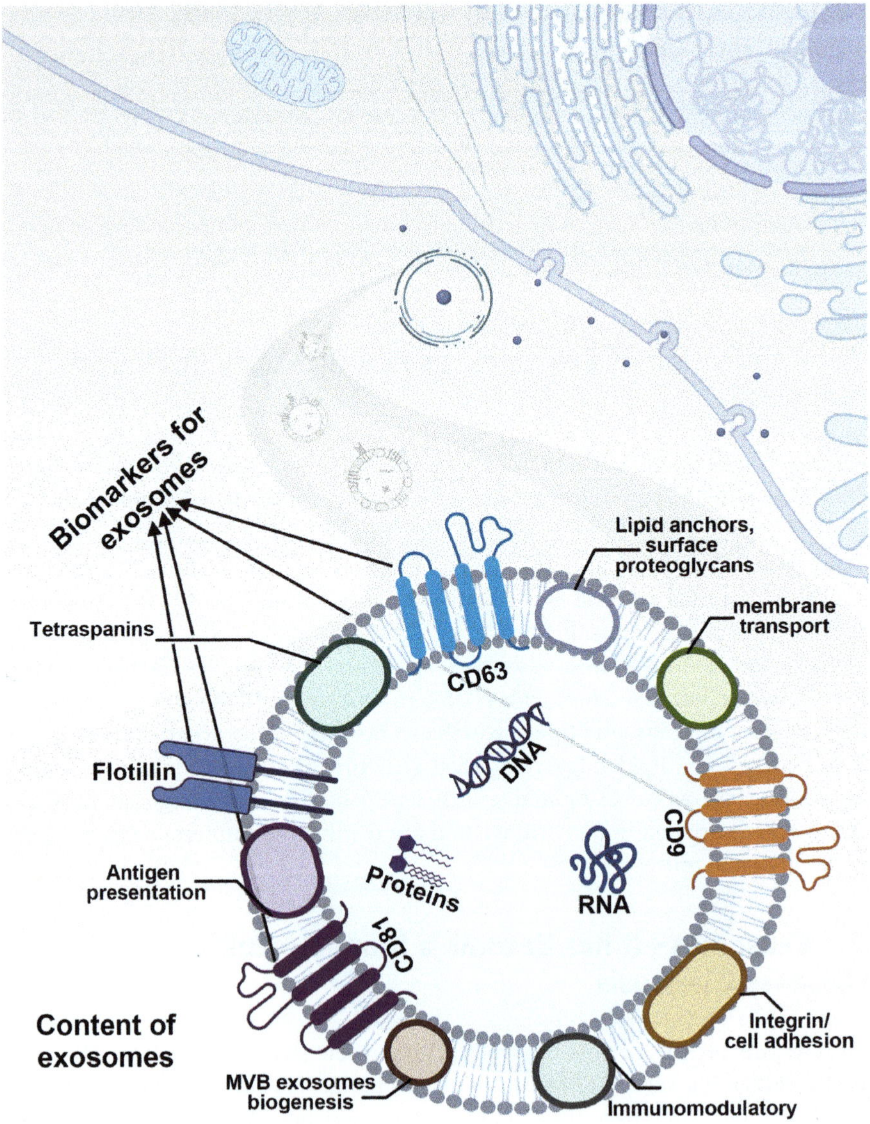

Figure 7.1 Biogenesis and identification of exosomes. Source: Kalluri and LeBleu [13]/ American Association for the Advancement of Science.

The biology of exosomes in disease is still emerging, and the number of studies addressing their utility in the diagnosis and treatment of various pathologies has increased substantially. It is worth highlighting that the exosomes present various advantages over other detection modalities, especially surface proteins on exosomes also facilitate their immune capture and enrichment. Diseases that have been the focus of the diagnostic application of exosomes include CVDs [14, 15], diseases affecting the central nervous system [16], and cancers [17, 18]. This effort is rapidly expanding to other diseases involving the liver [19], kidney [20], and lung [21].

Fluid and extracellular constituents such as proteins, lipids, metabolites, small molecules, and ions can enter cells, along with cell surface proteins, through endocytosis and plasma membrane invagination. The resulting plasma membrane bud formation in the luminal side of the cell presents with outside-in plasma membrane orientation. This budding process leads to the formation of endosomes (ESEs) or possible fusion of the bud with ESEs performed by the constituents of the endoplasmic reticulum (ER), trans-Golgi network (TGN), and mitochondria. The ESEs could also fuse with the ER and TGN, possibly explaining how the endocytic cargo reaches them. Some of the ESEs can therefore contain membrane and luminal constituents that can represent diverse origins. ESEs give rise to LSEs. The second invagination in the LSE leads to the generation of ILVs, and this step can lead to further modification of the cargo of the future exosomes, with cytoplasmic constituents entering the newly forming ILV. As part of the formation of ILVs, proteins (that were originally on the cell surface) could be distinctly distributed among ILVs. Depending on the invagination volume, the process could give rise to ILVs of different sizes with distinct content. LSEs give rise to MVBs with a defined collection of ILVs (future exosomes). MVBs can fuse with autophagosomes, and ultimately the contents can undergo degradation in the lysosomes. The degradation products could be recycled by the cells. MVBs can also directly fuse with lysosomes for degradation. MVBs that do not follow this trajectory can be transported to the plasma membrane through the cytoskeletal and microtubule network of the cell, and dock on the luminal side of the plasma membrane with the help of MVB-docking proteins. Exocytosis follows and results in the release of the exosomes with a similar lipid bilayer orientation as the plasma membrane. Several proteins are implicated in exosome biogenesis and include Rab GTPases, ESCRT proteins (see text), as well as others that are also used as markers for exosomes (CD9, CD81, CD63, flotillin, TSG101, ceramide, and Alix). Exosome surface proteins include tetraspanins, integrins, immunomodulatory proteins, and more. Exosomes can contain different types of cell surface proteins, intracellular proteins, RNA, DNA, amino acids, and metabolites.

Exosomal contents offer a rich source of potential biomarkers and may reflect the underlying molecular mechanisms of disease [22]. miR-7-1-3p and miR-211-5p have been shown to be markers of atherosclerosis, and in the case of miR-7-1-3p, incriminated in the pathogenesis of ischemic cardiac disease [15, 23, 24]. This demonstrates that an exosomal miRNA would be used as an auxiliary diagnostic marker for cardiac sarcoidosis (CS) [25]. The total circulating exosome transcriptome in relapsing-remitting multiple sclerosis (RRMS) patients and healthy controls (HC) were contrasted: four circulating exosomal sequences within the miRNA category were differentially expressed in RRMS patients versus HC: hsa-miR-122-5p, hsa-miR-196b-5p, hsa-miR-301a-3p, and hsa-miR-532-5p. Serum exosomal expression of these miRNAs was significantly decreased during relapse in RRMS [26]. There is, then, reason to believe that exosomal miRNAs might represent a useful biomarker to distinguish multiple sclerosis relapse.

7.2.1 Exosomes Could Be Used as Prognostic and Diagnostic Biomarkers in Cancer

Cancer is one of the important causes of death worldwide. Therefore, early detection of cancer is vital for increasing survival and effective patient management [27, 28]. Tumor biopsy is an invasive technique that is commonly used for the diagnosis of cancers. This technique is very high risk for patients and may threaten their lives. In addition, biopsies have potentially harmful effects stimulating cancer progression and metastasis [29]. Thus, finding blood biomarkers with high specificity and sensitivity can be a useful potential approach for noninvasive cancer diagnosis. In these regards, exosomal microRNAs (miRNAs) have attracted the attention of most of the researchers.

Some studies have suggested that small amounts of DNA can be found in exosomes, and that this DNA can be of value in detecting cancer-associated mutations in serum exosomes [30–33]. Specific miRNAs or groups of miRNAs in exosomes may provide diagnostic or prognostic potential in the detection of cancer [34]. Just after Valadi's discovery in 2007 [35], various studies have been performed in order to characterize exosomal miRNA as diagnostic biomarkers for cancers. In 2008, Taylor and Gercel-Taylor reported that eight miRNAs, including miR-21, miR-141, miR-200a, miR-200c, miR-200b, miR-203, miR-205, and miR-214, previously demonstrated as diagnostic markers for ovarian cancer, were also present in serum exosomes isolated from the ovarian cancer patients [36]. In 2009, Rabinowits and colleagues carried out a miRNA profiling analysis on tumor biopsy specimens, exosomes isolated from lung adenocarcinoma patients, and control subjects. They found a similar miRNA profile between exosomes and tumor biopsy samples from lung cancer patients, both significantly different from those detected in control subjects [37]. In addition, the exosomal miRNAs miR-378a, miR-379, miR-139-5p, and miR-200-5p are possible markers to discriminate tumors from normal samples in lung adenocarcinoma. It suggested that circulating exosomal miRNAs could potentially be used as a screening marker for lung adenocarcinoma. In esophageal squamous cell cancer (ESCC), the expression level of exosomal miR-21 from patients with ESCC is dramatically increased, compared to the patients with benign disease. Furthermore, exosomal miR-21 level is correlated with advanced tumor stages, lymph node involvement, and metastasis [38]. The use of exosomal miRNAs as diagnostic biomarkers has also been evaluated in other fluids, such as urine samples. For example, Armstrong et al. demonstrate that the expression of miR-21 and miR-4454 is significantly upregulated in the urinary exosomes of bladder cancer patients [39]. Furthermore, it has been reported that the expression level of exosomal miR-21-5p, miR-574-3p,

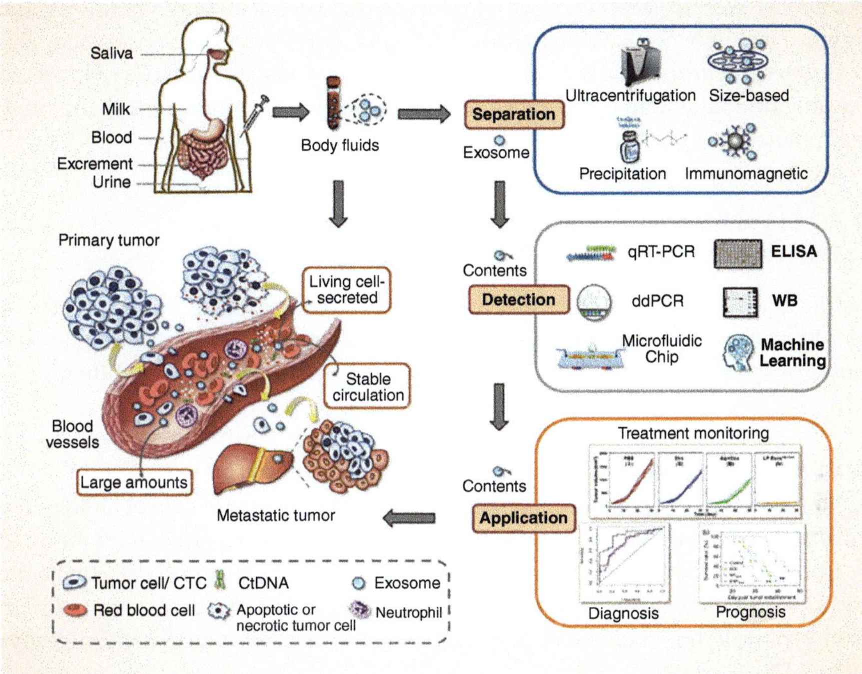

Figure 7.2 Exosomes as a new target for liquid biopsy. Source: Yu et al. [12]/ Springer Nature / CC BY 4.0.

and miR-141-5p is significantly upregulated in the urine of prostate cancer patients [40].

These findings have led to the idea that analyzing the circulating exosomes and their derived cargoes may provide new opportunities for cancer liquid biopsy (Figure 7.2), highlighting the potential of exosomes as biomarkers for cancer diagnosis, progression monitoring, and prognosis prediction.

Identification of the contents in exosomes provides more information on specific biomarkers than traditional tumor biomarker assays. Monitoring tumor exosome contents for specific biomarkers would be a convenient noninvasive method for diagnosis, and a way to follow disease progression and treatment [41]. Given the universality and diversity of their functional roles, exosomes (the sensitive and specific biomarkers) are useful for diagnosing the early stage of cancers and predicting an unfavorable prognosis, which can be obtained from the blood, urine, cerebrospinal fluid (CSF), or any other bodily fluid. Although increasing amounts of preclinical research have been reported in recent years, the studies addressing their utility in the diagnosis and treatment of various pathologies (clinical human trials) have yet to be confirmed. The clinical translation of exosomes is of primary importance; hence, accelerating exosomes cancer screening industrialization is of great significance.

Exosomes are enriched in body fluids and are critically involved in tumorigenesis, tumor progression, and metastasis. Compared with circulating tumor cells (CTCs) and circulating tumor DNA (ctDNA), exosomes show superior characteristics such as living cell-secreted vesicles, large amounts, and stable circulation. Traditional and advanced technologies have been used to separate exosomes from various body fluids and to detect exosomal cargoes. The detection of specific molecules of exosome may provide a new strategy for cancer diagnosis, progression monitoring, and prognosis prediction.

7.2.2 Exosomes Biopsy Strategies were Proposed to Target the Different Cancers

Identification and detailed analysis of cancer-derived exosomes may be useful in identifying tumor-preferred metastatic sites. Accurate knowledge of exosomal data transfer may open new perspectives in tumor diagnosis—monitoring of therapy in the future.

7.2.2.1 Pancreatic Cancer

Raghu Kalluri (the founder of *Codiak BioSciences*; the director of cancer biology, the University of Texas, M. D. Anderson Cancer Center (MDACC)) et al. have reported the utility of GPC1 (glypican 1)-positive exosomes in the diagnosis of pancreatic cancer, with GPC1 being enriched in cancer cell-derived exosomes compared to healthy controls; this could serve as an early detection tool for tumors in the digestive system [42].

7.2.2.2 Gastric Cancer

Sixty patients with gastric cancer (GC) and 30 patients with Chronic Gastritis (disease control group) admitted to The First Affiliated Hospital of Henan University from October 2019 to December 2020 were selected. Meanwhile, 30 healthy subjects (healthy control group) who underwent physical examination were also enrolled. Then, the expression difference of plasma exosomal hsa_circ_0064910 in GC patients before and after the operation was analyzed, and its relationship with clinicopathological features of GC patients was also investigated. All these data demonstrated that exosomal hsa_circ_0064910 is highly expressed in GC patients and might be a potential noninvasive biomarker for the auxiliary diagnosis of GC [43].

7.2.2.3 Lung Cancer

Some researchers revealed that some humoral-derived substances, such as exosomal miRNAs, are considered the most potential biomarkers for diagnosing early lung cancer in body fluids because of their stability, accessibility, and specificity. For example, Sun, Z. et al. reported that miR-96 has a significant diagnostic value in patients suffering from lung cancer [44]. Another study reported that miR-21 and miR-4257 expression were significantly correlated with recurrence and poor survival in lung cancer patients [45]. Moreover, four LUAD-specific miRNAs (miR-181-5p, miR-30a-3p, miR-30e-3p, and miR-320b) have been crucially produced as ideal biomarkers for distinguishing adenocarcinoma from squamous cell carcinoma in NSCLC with the area under the curve values of 0.936 [46, 47]. A study reported that exosomal MiR-205 expression separates squamous from nonsquamous NSCLCs even in poorly differentiated tumors; this made this exosome to be considered as a potential biomarker for lung cancer [48]. Besides, a study conducted by Wen Gao et al. revealed that the deregulated expression of miR-21, miR-143, and miR-181a in non-small cell lung cancer is linked to clinical pathology or patient prognosis. They concluded that the three exosomes could act as a novel diagnostic or prognostic biomarker for NSCLC [49].

7.2.2.4 Breast Cancer

The droplet digital ExoELISA (from Lei Zheng's research team, Southern Medical University) for exosome quantification has the potential to distinguish the target protein expression level on single exosomes through the fluorescence signal level in droplets. The high specificity was also demonstrated by quantifying the exosomes with target GPC-1 biomarkers from a variety of exosome subpopulation protein biomarkers. They successfully used this method for the absolute quantification of exosomes in serum samples from breast cancer patients, manifesting the prospective clinical value of the droplet digital ExoELISA method that may propel the discovery of cancer exosomal biomarkers [50].

7.2.2.5 Liver Cancer

Won Sohn et al. found that the levels of miR-18a, miR-221, miR-222, and miR-224 in exosomes of patients with HCC (Hepatocellular carcinoma) were lower than those in patients with HBV (Hepatitis B virus), indicating that these three miRNAs could be used as novel serum markers for detecting HCC [51]. Apart from miRNAs, lncRNAs are also used as biomarkers for clinical diagnosis of HCC [52]. Xiang Ma et al. discovered in an experiment that exosomes mediate the regulation of lncRNAs X-inactive-specific transcript in the expression of blood cells, and indicate that Xist expressed by mononuclear cells and granulocytes might act as valuable biomarkers in the diagnosis of female HCC patients [53]. Another study also reported that miR-30d, miR-140, and miR-29b showed significance in the survival of patients with liver cancer. Therefore, these exosomal miRNAs act as prognostic biomarkers for liver cancer and guide in the treatment of advanced liver cancer [54]. Another experiment also supported that exosomal lncRNA acts as a prognostic factor in HCC, and lncRNA (LINC00161) significantly upregulates in HCC patients when compared to normal patients, which is well stabilized and specific [55]. In addition to lncRNA, circPTGR1 (a circRNA) is particularly expressed in exosomes of 97 L and LM3 cells, and is increased in the serum exosomes of HCC patients. Wang G and his collaborator indicated that it could be used for clinical staging and prognosis [56].

7.2.2.6 Ovarian Cancer

In clinical practice, patients with ovarian cancer often suffer from venous thromboembolism (VTE) due to the aberrant activation of platelet and coagulation dysfunction [57]. Most of them remain in a hypercoagulable state, and the status of coagulation has emerged as an indicator of EOC. As shown in Table III, among the 50 DEGs, 10 genes participated in the complement and coagulation cascade, namely coagulation factor XIII A chain (F13A1), coagulation factor IX (F9), serpin family A member 1 (SERPINA1), FGA, FGB, FGG, complement C9 (C9), complement component 4 binding protein alpha (C4BPA), complement C8 alpha chain (C8A), and complement component 4 binding protein beta (C4BPB), the majority of which were upregulated in patients with ovarian cancer, suggesting that the overexpression of these exosomal genes mediates the coagulation cascade as well as platelet activation, and induces blood clotting. In total, four genes (LBP, FGG, FGA, and GSN) in exosomes derived from the 80 patient plasma samples were eventually selected as the potential diagnostic and prognostic biomarkers for the reason that they were all involved in two categories of function enrichment, namely coagulation and apoptosis-related pathways [58].

7.2.2.7 Melanoma

Cutaneous melanoma is an aggressive cancer type derived from melanocytes, and its incidence has rapidly increased worldwide [59]. The transmission of malignant tumor exosomes to normal melanocytes could confer vicious alteration to normal melanocytes. It is well validated that melanoma exosomes play a pivotal role in fueling the growth, metastasis, immune escape, and even drug resistance of melanoma by transferring carcinogenic nucleic acids and proteins [60]. Importantly, normal melanocytes acquire their invasiveness through melanoma exosome transported molecules. Melanoma exosomes support tumor progression by promoting angiogenesis, modulating the immune system, and remodeling tumor parenchyma [61]. Clinically, the exosomes isolated from cancer patients with metastasis or relapse, or in the advanced stage are very characteristic [62]. That means the content profile of exosomes may reflect the state of the cells that release them, providing valuable biological information on the parent cells [63]. These circulating exosomes represent important diagnostic and prognostic markers to capture the dynamic information of the tumor state. Thus, the qualitative and quantitative detection of exosomes in melanoma tissues and serum/plasma has aroused widespread concern in the field of diagnostic minimally invasive screening in liquid biopsy [64].

7.2.2.8 Colon Cancer

Colorectal cancer (CRC) ranks third in common malignancies and is the second leading cause of death concerning cancer [65]. Blood and face samples from CRC patients and healthy donors were collected at Tianjin People's Hospital. The research team identified two biomarkers on fEVs for CRC detection—CD147 and A33. They discovered that CD147 and A33 on fEVs could distinguish CRC patients from healthy donors with AUC values of 0.903 and 0.904, respectively. The biomarkers CD147 and A33 on fEVs outperform those in plasma EVs and serum CEA regarding diagnostic accuracy, compliance, and simplicity. This study shows that fEVs can be used as new biomarkers for CRC diagnosis and prognosis with high clinical sensitivity and sensitivity, providing new opportunities for mass CRC screening in an entirely noninvasive manner.

7.2.2.9 Glioma

Glioma is the most common malignant primary brain tumor in the central nervous system (CNS) [66]. Yanlian Yang et al. present a latex bead-assisted flow cytometry method for the detection and molecular analysis of EVs, and examine the clinical value of this method in diagnosing and predicting malignancy in glioma patients [67]. The study was approved by First Affiliated Hospital of Chongqing Medical University (2017-032). Twenty-three glioma patients and 12 healthy donors who had no sign of serious disease and no surgery within the past 12 years were recruited on the basis of an institutional board-approved protocol. All patients signed the informed consent form. Moreover, we found that (epidermal growth factor receptor) EGFR in serum EVs can accurately differentiate high-grade and low-grade glioma patients, and EGFR in EVs positively correlates with ki-67 labeling index (LI), which represents the percentage of ki-67-positive cells in the tumor tissue assessed by immunohistochemistry that has been shown to be associated with the high malignancy and poor outcome of tumors [68–70], indicating that EGFR in serum EVs can reflect the malignancy of glioma. We also demonstrate the efficacy of NLGN3 and PTTG1 mRNA in serum EVs to diagnose glioma patients. These results demonstrate the clinical significance of serum EVs in the diagnosis and prognosis prediction of glioma, and will be beneficial for glioma cancer management.

Due to its ease of access and stability, exosomes have been proposed as a novel, minimally invasive tool for cancer diagnosis. In further study and clinical application of liquid biopsy, the combination of various contents of exosomes as biomarkers can provide a more effective guarantee for the accuracy of cancer diagnosis. To date, most studies on cancer therapy response assessment remain at the level of cancer cells *in vitro*, and more clinical trials are needed to validate the clinical capacity of these biomarkers (Table 7.1).

The recent clinical studies on exosomes as biomarkers for early diagnosis include cohort scale, source of body fluid, functional targets of exosomes, exosome extraction methods, target isolation, and detection methods [72].

7.2.3 Exosomes in Clinical Trial for Cancer Biopsy

The above reports on exosomes from various cancers that we have listed were based on clinical presentations or laboratory tests in clinical practice. All these samples collected are from human serum, plasma, urine, resolution, tissue fluid, and other body fluids, which demonstrated once again that exosomes are indeed promising for future applications in cancer detection. That's one small step for medical applications; however, intense research and numerous clinical trials are urgently needed.

Exosome-based liquid biopsy has been tested in clinical trials, and several of them have been approved and reached the market. In 2016, Exosome Diagnostics proposed the first exosome-based liquid biopsy in the world—ExoDx™ Lung (ALK)—for the isolation and analysis of exosomal RNA from blood samples. At a CLIA-certified laboratory, ExoDx Lung (ALK) was proved to be an accurate, real-time tool to detect EML4-ALK mutations in NSCLC patients with 88% sensitivity and 100% specificity, which provides a more direct and sensitive method

Table 7.1 Exosome as potential predictive biomarkers in different cancers.

Applications	No. of patient	Source	Volume of body fluid	Targets	Exosome extraction	Extraction method	Detection method	References
Pancreatic cancer [42, 80, 81]	263	Plasma	0.9–1.5 ml	*KRAS*	Ultracentrifugation	MagAttract High Molecular Weight DNA kit	ddPCR	(Allenson et al., 2017)
	221	Serum	250 μl	GPC1	Ultracentrifugation + sucrose gradient centrifugation	Affinity capture	Flow cytometry analysis	(Melo et al., 2015)
	85	Serum	–	CKAP4	PS Capture Exosome ELISA Kit	PS Capture Exosome ELISA Kit	PS Capture Exosome ELISA Kit	(Kimura et al., 2019)
Colorectal cancer [82–86]	140	Serum	300 μl	lncRNA	Ultracentrifugation	TRIzol	qPCR	(Dong et al., 2016)
	40	Serum	250 μl	miRNA	ExoQuick-TC™ Exosome Precipitation Solution kit	miRNeasy Serum/Plasma kit	qRT–PCR	(Zhao et al., 2017)
	102	Tissue homogenate	–	GPC1	ExoCap™ Exosome Isolation, Enrichment kit	Affinity capture	Flow cytometry analysis	(Li et al., 2017)
		Plasma		miR-96-5p, miR-149		TRIzol	qPCR	
	116	Serum	250 μl	CEA	ExoQuick™ Exosome Precipitation Solution	ELISA	ELISA	(Yokoyama et al., 2017)
	124	Plasma	500 μl	Copine III	Ultracentrifugation	PIPA buffer	ELISA	(Sun et al., 2019)

Lung cancer [71,87–89]	210	Plasma	1–2 ml	*EGFR T790M*	Ultracentrifugation	ExoLution Plus platform	Allele-specific qPCR assay	(Castellanos-Rizaldos et al., 2018)
	105	Plasma BALF	–	miRNAs	Ultracentrifugation	mirVanaTM miRNA isolation kit	qPCR	(Rodríguez et al., 2014)
	581	Plasma	–	Proteins	Extracellular vesicle array	Extracellular vesicle array	Extracellular vesicle array	(Sandfeld-Paulsen et al., 2016)
	171	Serum	–	Proteins	Ultracentrifugation	Lysis buffer	Immunoblotting	(Niu et al., 2019)
Breast cancer [90–93]	32	Plasma	250 µl	miR-21, miR-1246	Exoquick-TC™ reagent	TRIzol	qRT–PCR	(Hannafon et al., 2016)
	38	Serum	–	miR-105	Ultracentrifugation	TRIzol	qRT–PCR	(Zhou et al., 2014)
	240	Serum or Plasma	500 µl	CD82	ExoQuick™ exosome precipitation solution	Strong RIPA lysate	ELISA/western blot	(X. Wang et al., 2019)
	44	Plasma	5.5 ml	Phospho-proteins	Ultracentrifugation	–	LC–MS/MS	(Chen et al., 2017)
Renal cell carcinoma [94]	52	Urine	–	Proteins	Differential centrifuga-tion + density gradient ultracentrifuga-tion/ultrafiltration	–	LC–MS/MS	(Raimondo et al., 2013)
Bladder cancer [95]	69	Urine	38.5 ml	miRNAs	Differential centrifugation	miRNeasy Mini Kit/RNA MS2/Clean-up Kit	miRNA microarray	(Matsuzaki et al., 2017)
Cholangio-carcinoma [96]	134	Serum	1 ml	Proteins	Ultracentrifugation	–	MS	(Arbelaiz et al., 2017)

to detect gene fusions than cfDNA. In addition, the ExoDx Prostate IntelliScore (EPI) has been certified by the FDA. Based on the detection of ERG, PCA3, and SPDEF RNA in exosomes, EPI provides a risk score to predict whether a patient with PSA from 2 to 10 ng/mL is likely to develop higher-grade prostate cancer [73]. According to ExoDx, 93% of sensitivity was achieved in prospective studies, and 26% of unnecessary needle biopsies were avoided when the EPI threshold was set at 15.6 [74]. Three independent, prospective, and multicenter clinical trials declared that EPI outperformed the standard of care, and could be used to assist in the early diagnosis of prostate cancer and eliminate unnecessary prostate biopsy [75]. Moreover, MedOncAlyzer 170 is a newly developed liquid biopsy system capable of detecting both exosomal RNA and ctDNA in a single trial. It can identify significant and functional mutations in multiple cancer types from small volumes (0.5 ml) of patient blood or plasma. Due to the unique formation manner of exosome and ctDNA, MedOncAlyzer 170 is accurate and highly sensitive to detect mutation at all stages of cancer progression and treatment. Although the clinical application value has been verified, larger clinical samples, populations, and trials are still needed to confirm the role of exosome-based liquid biopsy in cancer diagnosis and treatment.

7.3 The Commercial Application of Exosomes

7.3.1 Tumor Therapy

Exosomes derived from tumor cells typically contain some tumor antigens that can activate antigen-presenting cells. Therefore, exosomes can be used for the development of tumor vaccines, and some studies have found that such tumor vaccines have good feasibility and safety. According to the medical magic cube data, the Chimeric exomeric tumor vaccine (chimeric exocrine tumor vaccine) from Zhangjiang, Fudan, is currently in phase I clinical research for the treatment of bladder cancer.

Immune-cell-derived extracellular vesicles can mediate and regulate the body's immune response to tumors. The potential of extracellular vesicles in tumor immunotherapy has been widely studied. Dendritic cell (DC)-derived extracellular vesicles contain a large number of MHC–peptide complexes and immune stimulatory molecules, which can activate relevant T cells and mediate antitumor responses in the body, showing good antitumor efficacy in multiple clinical trials. EVs derived from CAR-T cells express high levels of perforin and granzyme B and CAR, inducing the death of tumor cells expressing the same antigen. Due to their small size, extracellular vesicles can cross biological barriers and effectively penetrate solid tumors [76].

7.3.2 Lung Infection and ARDS Treatment

In March 2021, the United States FDA approved Direct Biologics' application for ExoFlo™, a stem cell exosome medication, to enter Phase I/II clinical trials. ExoFlo is a cell-derived extracellular vesicle biological product known as exosomes that are isolated from human bone marrow mesenchymal stem cells. Its initial indications included treating moderate-to-severe COVID-19 infections and acute respiratory distress syndrome (ARDS). According to Direct Biologics, the Phase II clinical trials for ExoFlo showed no adverse events (AE) or serious adverse events (SAEs). In the ExoFlo-15ml group, the mortality rate was significantly lower than the placebo group, reducing the absolute and relative risks by 30.8% and 61.6%, respectively, in respiratory failure patients aged 18–65, and by 41.9% and 57.7%, respectively, in ARDS patients aged 18–65.

In July 2022, ExoFlo entered Phase III clinical trials. By January 2023, ExoFlo had successfully completed the enrollment of 400 patients in clinical Phase III trials, achieving a significant milestone in the medical community, as it became the first-ever application of stem cell exosomes approved by the FDA for disease treatment. In addition to its indications in COVID-19 and ARDS treatment, ExoFlo is currently undergoing clinical Phase I trials for other indications, including Crohn's disease, irritable bowel syndrome, ulcerative colitis, organ transplant rejection, and lower back pain.

Direct Biologics believes that ExoFlo's success is due to the company's cGMP amplification and sterile processing, which has resulted in ultrapure and sterile exosomes. With over 100 billion exosomes per milliliter, ExoFlo has demonstrated excellent efficacy, and perfect quality control has ensured that every vial of exosomes has consistency in its quality (Figure 7.3).

The haMPC-Exos of Xibiman is being used in the IIT study for pulmonary infections, with 60 volunteers enrolled and currently being carried out at Ruijin Hospital (NCT04544215). Experimental studies related to lung injury indicate that extracellular vesicles from stem cells have a positive therapeutic effect. Vitti Labs' EV-Pure extracellular vesicle drug is being used for the treatment of pulmonary fibrosis and primary ovarian insufficiency in females, and entered phase I clinical trials in October 2022 (NCT05387239).

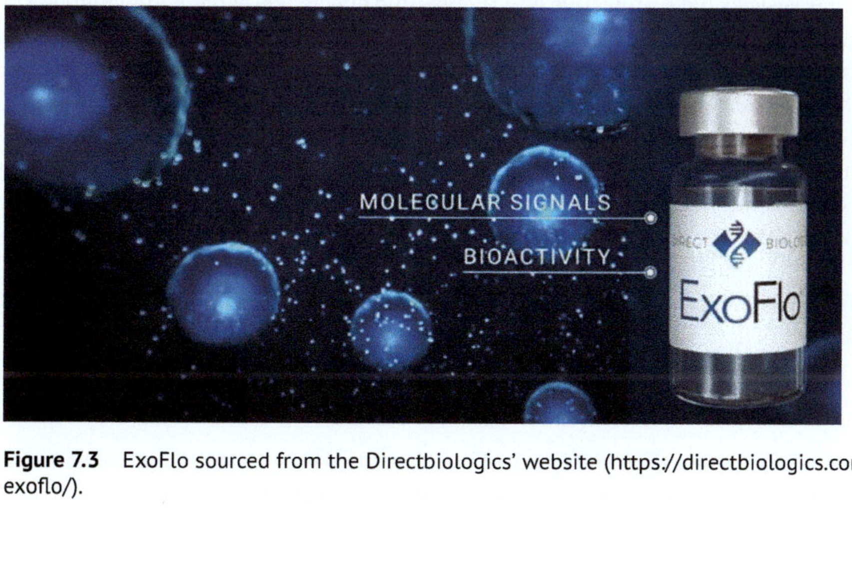

Figure 7.3 ExoFlo sourced from the Directbiologics' website (https://directbiologics.com/exoflo/).

7.3.3 Cardiovascular Disease Treatment

The heart protection mechanism of mesenchymal stem cell-derived exosomes has become a hot research topic. Increasing evidence shows that mesenchymal stem cell-derived exosomes have similar heart protection functions to mesenchymal stem cells, significantly improving the survival rate of myocardial cells, protecting against myocardial cell damage, promoting neoangiogenesis, improving heart function, anti-inflammation, antimyocardial cell apoptosis, and promoting myocardial cell regeneration. Exosomes have good stability, and cause fewer mutations and immune rejection reactions in recipient cells, making them a promising method for treating cardiovascular diseases.

Mesenchymal stem cell-derived exosomes have great potential in the treatment of cardiovascular diseases such as myocardial infarction and heart failure. A review published in STEM CELLS journal discusses the methods of promoting angiogenesis and enhancing their bioactivity to improve heart tissue repair. In April 2022, a clinical trial initiated by the Tehran University of Medical Sciences to evaluate the efficacy of mitochondria and MSC-derived exosomes for the treatment of myocardial infarction was officially launched, with 20 volunteers recruited to participate (NCT05669144).

Mesenchymal stem cell-derived exosomes have good efficacy in ischemic heart disease. A study published in BioMed Res found that injection of mesenchymal stem cell-derived exosomes inhibited fibrosis and inflammation in a rat model of ischemic heart disease, and significantly improved heart function. In September 2021, the latest clinical study (NCT03384433) from Isfahan University of Medical Sciences explored the clinical therapeutic effect of mesenchymal stem cell-derived exosomes carrying miR-124 on ischemic heart disease through five patients.

7.3.4 Liver and Kidney Injury Treatment

Liver disease-related experimental studies have shown that stem cell-derived extracellular vesicles have protective and therapeutic effects. Research conducted by Du et al. has demonstrated that stem cell-derived extracellular vesicles protect the liver from hepatic ischemia/reperfusion injury by activating the SPhks and s1p signaling pathways while also promoting cell proliferation. Similarly, Li et al. found that stem cell-derived extracellular vesicles improve liver fibrosis by suppressing the migration of hepatic stellate cells and the production of collagen. Experimental studies on kidney injury have shown that extracellular vesicles carry miRNA and proteins that can alleviate renal ischemia/reperfusion injury, reduce cell apoptosis, promote cell proliferation, and repair kidney damage. In April 2022, the extracellular vesicle drug ILB-202 from ILIAS Biologics entered clinical phase I targeting NF-κB for the treatment of acute kidney injury.

7.3.5 Ophthalmology Treatment

Extracellular vesicles can stably exist in ocular fluids such as tears, aqueous humor, vitreous humor, and blood, playing an important role in regulating ocular cell migration, proliferation, apoptosis, immune response, and angiogenesis. They are closely related to various eye diseases, including corneal injury, diabetic retinopathy, age-related cataracts, glaucoma, age-related macular degeneration, and choroidal melanoma.

Wharton jelly derived mesenchymal stem cell exosomes for the treatment of retinal pigment epithelial dystrophy are in phase II/III clinical trials (NCT05413148) at Erciyes University, with 135 volunteers enrolled since August 2022.

PSC-MSC-Exo, the exosomes of native mesenchymal stem cells, are undergoing phase II studies for the treatment of dry eye. Umbilical cord-derived mesenchymal stem cell-derived exosomes for the treatment of dry eye are also undergoing phase II studies at Sun Yat-sen University.

7.3.6 Cartilage Injury Treatment

Osteoarthritis is a common clinical condition characterized by joint cartilage damage, subchondral bone sclerosis, synovial inflammation, and so on. In recent years, numerous studies have demonstrated that stem cell-derived extracellular vesicles have anti-inflammatory, immune regulatory, tissue repair, and regeneration properties, which are a new treatment for osteoarthritis and cartilage injury. Extracellular vesicles from stem cells from various sources can protect and treat osteoarthritis. SF-MSC-EX is in phase II clinical trial for treating cartilage injury (NCT05261360), with 30 participants recruited in March 2022. CelliStem OA-sEV from Cells for Cells is in a phase I clinical trial for treating knee osteoarthritis. Autologous AdMSC-Exo treated for osteoarthritis research is undergoing IIT at West China Hospital of Sichuan University.

7.3.7 Other Treatments

Stem cell exosomes can be used to treat diseases such as stroke, diabetes, premature ovarian failure, Crohn's disease, irritable bowel syndrome, skin-related diseases, graft-versus-host disease, and wound healing.

7.3.8 Engineering of Exosome Delivery

Scientists construct engineered exosomes by using techniques such as endogenous or exogenous loading (1.4 engineering exosomes), which can deliver proteins, nucleic acids, and peptides to achieve innovative drugs and vaccines (category 1 new drugs), or small-molecule chemical drugs or effective ingredients in traditional Chinese medicine to achieve new uses for old drugs (category 2 new drugs/505b2). This revolutionary next-generation nanocarrier is expected to improve drug efficiency, reduce dosage frequency, enhance therapeutic effects with combined drug use, and minimize systemic exposure to side effects, globally. As of now, there are five engineered exosome drugs in clinical trials, with one in Investigator-Initiated Trials (IITs).

CovenD24, also known as EXO-CD24, was jointly developed by OBCTCD24 Ltd. and Elpen Pharmaceutical Co. Inc. for the treatment of COVID-19. It has undergone one Phase I clinical trial and four Phase II clinical trials (NCT04969172, EudraCT2021-002184-22, EudraCT2021-004259-17, NCT04902183, NCT04747574), with a total of 526 volunteers enrolled and positive results obtained. Over 90% of COVID-19 patients in the trials were cured and discharged within five days. CovenD24 is a medication that uses exosomes as carriers to contain CD24 proteins. CD24 primarily inhibits immune responses and cytokine release by binding to specific cell receptors, effectively combating COVID-19. However, CD24 has not yet been approved for use as a therapeutic target, and Merck's CD24-Fc fusion protein medication (MK-7110), acquired for $425 million in advance payments, has terminated its Phase III clinical trials for COVID-19 treatment and is currently undergoing clinical I/II trials for solid tumor indications. Therefore, choosing CD24 as the target for CovenD24 research and development carries significant risks.

ExoIL-12™ and ExoSTING™, developed by Codiak BioSciences, are two products based on the company's engineered exosome technology platform—engEx™ PLATFORM. They are also the most advanced engineered exosome drug delivery products to enter clinical trials worldwide. The platform uses PTGFRN and BASP1 as scaffold proteins, and installs targeting ligands or therapeutic molecules onto the surface PTGFRN protein or the interior BASP1 protein of exosomes, achieving the transport to target tissues and cells, and treatment of diseases (Figure 7.4).

ExoIL-12 is a therapeutic candidate drug that utilizes extracellular vesicles to display fully active IL-12 on the surface through the PTGRFN scaffold protein. Looking back at history, interleukin products have strong antitumor immune responses by activating T cells and NK cells, but due to the inability to sustain local retention and systemic exposure causing severe adverse events, multiple clinical trials have failed, especially in the past year of 2022. Therefore, adding the highly risky IL-12 in the new field of extracellular vesicles may bring significant research and development risks for Codiak. ExoIL-12 is indicated for cutaneous T-cell lymphoma (CTCL) and administered by intratumoral injection. In December 2020, Codiak announced that

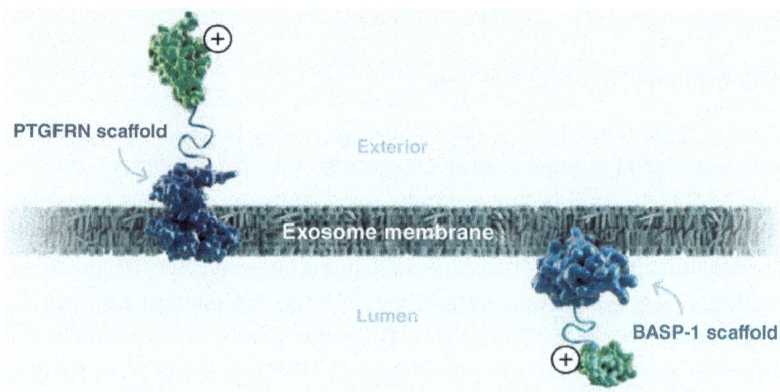

Figure 7.4 engEx™ PLATFORM sourced from the Codiak BioSciences.

the phase I clinical trial had achieved the primary endpoint, with good safety and tolerance demonstrated in all 55 healthy volunteers. No local or systemic adverse events related to treatment were reported, and no systemic exposure to IL-12 was detected. In June 2022, Codiak announced partial results of the CTCL trial, with two patients receiving over 20 injections of ExoIL-12 (6.0 μg) at multiple tumor lesions for more than six months without any grade 3 or higher treatment-related adverse events. Patient 001 showed partial response, with a 61% reduction in disease burden and 20–80% improvement in CAILS scores of all treated and untreated skin lesions. The treating physician deemed that further injections were unnecessary. Patient 002 received 20 injections, with a 43% reduction in disease burden and 30–50% improvement in CAILS scores of all treated and untreated lesions. While announcing the data results, Codiak also revealed plans to start a phase II trial in Q1 2023. However, in August 2022, the company abruptly announced the suspension of the clinical phase II trial of ExoIL-12 without disclosing the specific reasons.

ExoSTING is a product that incorporates STING agonists into the cavity of exosomes and expresses high levels of PTGFRN protein on the surface of exosomes to promote specific uptake of tumor antigen-presenting cells. This product is still administered by intratumoral injection to treat head and neck squamous cell carcinoma, triple-negative breast cancer, thyroid anaplastic carcinoma, and skin squamous cell carcinoma and completed a phase I/II clinical trial with 27 volunteers in September 2022. In all dosage groups, ExoSTING showed good tolerability, and no grade 3 or higher dose-limiting toxicity or treatment-emergent adverse events were observed, but treatment-related adverse events (TRSAEs) were observed in three patients: two with grade 2 cytokine release syndrome and one with grade 1 fever. Signs of antitumor activity were observed in injected and non-injected distant tumors. The data support the progression of ExoSTING to phase II clinical trials, especially in early disease, where the combination with immunotherapy could enhance efficacy. The company plans to start phase II clinical trials in the first quarter of 2023, but the company suddenly announced a suspension of the ExoSTING project in August 2022.

exoASO™-STAT6 is an antisense oligonucleotide (ASO) against STAT6 transcription factor loaded on the surface of exosomes, while expressing high levels of PTGFRN protein on the surface to promote selective uptake by M2-polarized tumor-associated macrophages. The indications of this product are advanced hepatocellular carcinoma (HCC), gastric cancer metastasized to the liver, and colorectal cancer metastasized to the liver, administered by intravenous injection, currently in clinical phase I with 30 volunteers enrolled, and no clinical data disclosed yet.

Milk is an expandable source of exosomes, containing abundant exosomes from various cell sources. It has been proven that milk exosomes (MEs) are evolutionarily conserved extracellular vesicles (EVs) with good stability and low immunogenicity, demonstrating excellent cross-species tolerance. Milk exosomes can survive under strong acidic conditions in the stomach and degradation conditions in the intestine, maintaining structural integrity, and exerting their function locally or after transportation to the circulatory system. The ability of milk exosomes to cross the gastrointestinal barrier makes them a promising oral delivery tool, and has been widely accepted as a promising oral carrier for transporting various biological molecules and chemotherapy drugs. On 11 March 2023, Tingo Exosomes Technology Co., Ltd. in Tianjin developed a new oral delivery system for mRNA-milk exosomes, and the relevant research results were published in *Vaccines*. Milk exosomes can protect mRNA from digestive enzymes and pH changes. Based on these characteristics, the R&D team of Tingo Exosomes Technology Co., Ltd. developed an efficient mRNA-milk exosome oral vaccine, which achieved the initial delivery of SARS-CoV-2 RBD mRNA.

The "EXC Exosome Bio-targeting Drug Delivery Platform" independently developed by Tianjin Exosome Technology has achieved exosome targeting delivery of Her2 antibodies, with a positivity rate of 76% for Her2 antibody loading on exosomes. The exosomes loaded with antibodies were taken up by Her2-positive cells 6.3 times and 174.5 times higher than controls under 37 and 4 °C, respectively (related reading: Tianjin Exosome achieves exosome exogenous loading targeting delivery). In addition to antibody coupling, Tianjin Exosome Technology used the small-molecule artemisinin–hemisuccinate as a model drug to achieve a loading rate and encapsulation rate of over 90%, with each exosome being able to load more than 7 x 10E5 artemisinin–hemisuccinate molecules, achieving milligram-level drug loading. The company used the EXC platform to design and prepare the first EDC (exosome–drug conjugation) template drug, chemically coupling antibodies, and small molecules on exosome surfaces, breaking through the bottleneck of low drug loading rate and limited safety of ADC drugs.

"Extracellular vesicles/extracellular vesicles from cells" carry highly bioactive substances such as proteins and active factors secreted by original cells that activate cell regeneration function. After penetrating into the skin, the vesicles release the nutrient substances wrapped inside, continuously nourish, adjust the cell microenvironment, enhance cell vitality, promote cell migration, promote collagen and elasticity to repair damage, regenerate cells, and restore youthful skin. These applications have great potential in the fields of medical beauty and skincare. Products from companies such as ExoCoBio in Korea (ASCE+™, EXOMAGE™ and CELLTWEET™), Baipao Medical (Fweiyi, Zhanweiya, Mile), Yuanxiang Biology (Ningguang E.R.T), Sanhe Medical (Aoyongli), Hanma Medical (Aobenyuan), Tianjin Extracellular Technology (Milk Extracellular Raw Materials+Y3), and Clarins (Radiance Prodige) have already been commercialized.

7.3.9 Skin Repair and Medical Skincare Products

In 2019, ExoCoBio published a scientific paper in *Stem Cell Research and Therapy*, proving for the first time that the extracellular vesicles derived from stem cells can significantly improve "atopic dermatitis," suppress various inflammation targets, and promote wound healing. Research has shown that stem cells have the characteristic of paracrine secretion, which can repair damaged tissue, promote cell growth, and regulate immune reactions. The extracellular vesicles secreted by them can reach distant damaged sites through most capillaries and can be

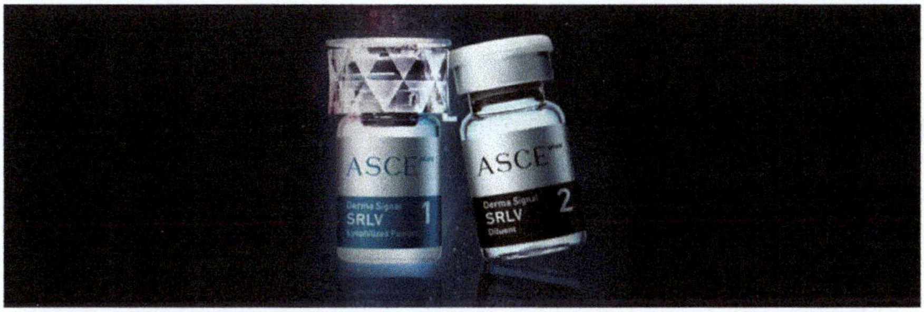

Figure 7.5 ASCEplus sourced from the ExoCoBio's website (http://www.exocobio.com/bizdemo87737/eng/03/01.php) .

combined with existing known active drugs to enhance the therapeutic effect, promote inflammation elimination, and wound healing.

"ASCE plus-SRLV" is the world's first product made by freeze-drying the extracellular vesicles (ASC-EXOSOME™) derived from stem cell culture medium. It features excellent "skin regeneration" and "anti-inflammatory" effects. The product consists of powder and liquid ampoules containing freeze-dried extracellular vesicles and various effective ingredients that are mixed and used together. ASCE plus extracellular vesicles can reduce 55% of inflammatory cytokines, promote seven times the synthesis of collagen, and increase three times the synthesis of elastic protein, achieving skin repair and regeneration, delaying aging, improving acne-prone skin and inflammation, and comprehensively improving skin problems. In June 2021, ExoCoBio and Biobijou company cooperated to enter the Chinese market (Figure 7.5).

One study, in which Jurlique participated and was published in the *Nature* journal, found that extracellular vesicles released by keratinocytes can regulate the pigmentation of melanocytes. These vesicles carry tiny RNAs that can interact with melanocytes to regulate pigmentation. Based on this research, Jurlique discovered smooth gold kiwi extract, which can whiten and fade blemishes by inhibiting the information contained in the extracellular vesicles. Currently, smooth gold kiwi extract is used in Jurlique skincare products, including as a main component in the whitening product—Radiant Serum.

French cosmetic raw material supplier Silab, in cooperation with France's National Institute of Health and Medical Research (INSERM), discovered in their research that fibroblast extracellular vesicles play a critical role in hair growth: The communication between extracellular vesicles between the dermis and hair follicles is a worthy biological pathway of attention in hair growth. Extracellular vesicles from fibroblasts can promote the secretion of Norrin (a ligand that can activate the beta-catenin pathway involved in hair growth). Based on this finding, Silab developed a natural hair growth ingredient called HAIRGENYL® yeast extract, which targets the "extracellular vesicle communication system between the dermis and hair follicles." In April 2022, Korean extracellular biotechnology company ExoCobio launched ASCE plus-HRLV—the world's first extracellular-based anti-hair loss and hair growth product.

7.4 Commercial Development of Exosomes

The extracellular vesicle industry is in its early stages and has not yet formed a complete industry chain pattern. The upstream of the industry chain consists of raw materials, equipment, and service suppliers; the midstream is composed of extracellular vesicle therapy drugs, medical aesthetic products, and diagnostic reagent suppliers; and the downstream is composed of circulation and application terminals.

The commercial application of the exosomes industry has attracted the attention of large multinational pharmaceutical companies, and the trend of industrialization cooperation between emerging biotechnology companies and large multinational pharmaceutical companies is gradually becoming clear. Large multinational pharmaceutical companies represented by Lilly, Takeda, and Jazz Pharmaceuticals have reached multiple high-value collaborations with companies with innovative exosome technology platforms and have successively signed several $1 billion level large-scale cooperative development agreements to jointly develop exosomes' targeted therapies.

Jazz Pharmaceuticals: In January 2019, Jazz Pharmaceuticals and Codiak reached a new drug research and development cooperation for hematological malignancies and solid tumors with extracellular drug delivery as the core. Jazz Pharmaceuticals has agreed to pay Codiak $56 million in research funding and up to $20 million in preclinical development milestone payments for the five new drug projects it collaborates with. If the project achieves the agreed milestones, an additional $200 million will be paid for each target. After the completion of Phase I/II clinical trials, Jazz Pharmaceuticals will be responsible for subsequent clinical trials, new drug applications, and commercial promotion.

Eli Lilly Pharmaceuticals: On 9 June 2020, Eli Lilly and Evox reached a cooperation agreement of 1.2 billion dollars to use its technology platform to develop engineered exosomes for delivering RNA interference drugs and antisense oligonucleotide drugs (ASO) to treat central nervous system disease.

Takeda Pharmaceuticals: In March 2020, Takeda also reached an $882 million partnership with Evox to use extracellular vesicles to deliver protein and mRNA nucleic acid drugs for the treatment of rare diseases, including type C Niemann Pick's disease. Evox will be responsible for promoting preclinical research and conducting clinical trial applications.

A multitude of companies have been established to exploit biotechnology development in exosomes. Founded in 2015, Codiak BioSciences is a pioneer biotechnology company developing exosome treatments for various diseases and is headquartered in Cambridge, Massachusetts. Codiak BioSciences has developed the engEx Platform to engineer exosomes to express and deliver therapeutic drug candidates. ExoSTING is a major and promising immune therapeutic candidate targeting cancer. Compared to free STING agonists, exoSTING is highly potent with minimal toxic potential. It is rarely affected by serum systemic cytokines, and preserves the vitality of effector T cells and antigen-presenting cells in tumors to maintain sustained immune protection. Recently, exoSTING is being developed for solid tumor therapy that activates the "STING" receptor in immune cells. Relevant clinical trials will be carried out in the first half of 2020 [77]. Exosome Diagnostics is a revolutionary developer of molecular diagnostics based on biological fluids, which was acquired by Bio-Techne last year. They are aimed at developing novel and precise exosome technology mainly in liquid biopsy of multiple cancers, including lung [71, 78] and prostate cancer [73, 74]. The ExoDx Prostate® (IntelliScore) (EPI) test is the star production of exosome diagnostics. This is a urine-based and completely noninvasive test designed to assist physicians in assessing whether an individual patient over 50 years old tested with 2–10 ng ml^{-1} prostate-specific antigen, which presents for a needle biopsy, is at greater risk for high-grade prostate cancer; therefore, the patient can avoid unnecessary biopsy and, instead, continue to follow up [74]. Moreover, the commercial exosome isolation kit ExoLution Plus Isolation Kit owned by Exosome Diagnostics is widely used in exosome research. Recently, Avalon GloboCare and its subsidiary Genexosome Technologies clarified the establishment and development of the first salivary-based exosome miRNA biomarker-miR-185 as a dual target for the diagnosis and treatment of oral cancer (FREEHOLD 2019: Avalon GloboCare and its Subsidiary Genexosome Technologies Announce Discovery and Development of World's First Saliva-Based Exosomal Biomarker "miR-185" as Dual Diagnostic and Therapeutic Target for Oral Cancer. Oral leukoplakia, with a prevalence of 2% affecting the worldwide population, is a pre-cancerous lesion that confers an increased risk for the development of oral cancer. Previously, there had been no reliable methodology to predict the progression from oral leukoplakia to malignant oral cancer. In collaboration with Beijing Stomatological Hospital affiliated with the Capital Medical University in China, Avalon and Genexosome Technologies have completed a clinical study and revealed miR185 as a novel saliva-based exosomal biomarker with strong correlation and predictive value for malignant transformation from oral leukoplakia to oral cancer. In a subsequent study, the companies further demonstrated that topical application of exosomes released from genetically modified human stem cells with increased expression levels of miR-185 can remarkably deter the progression of pre-malignant oral leukoplakia to form oral cancer. This study has been accepted as a poster presentation at the upcoming 2019 Annual Meeting of the International Society of Extracellular Vesicles (ISEV). The Company also announced the publication of a PCT patent application covering a method for preventing and treating oral cancer with extracellular vesicles (exosomes) carrying miR-185 (Publication No. WO 2018/205978). This PCT application allows Avalon and Genexosome Technologies to file patent applications and seek protection in most major national and regional markets throughout the world. http://puretechhealth.com/news/23-press-releases/1029-puretech-health-announces-collaboration-with-roche-to-advance-technology-for-oral-administration-of-antisense-oligonucleotides (Accessed 20 Jul 2018).) collaborated with Roche to impel the advancement of technology for oral administration of antisense oligonucleotides with PureTech's milk exosome-based technology [79] to transform conventional intravenous injection therapy for improved efficacy and reduced toxicity.

7.4.1 Analysis of Representative Companies of Exosomes

According to a 2018 research report by Grand View Research, as a drug delivery system alone, the exosomes market size is expected to reach $2.28 billion by 2030. Currently, multiple giants worldwide, including Roche, Takeda, and Lilly, are actively expanding the exosomes market.

According to the "2022–2040 Exosomes Therapy Market Report" released by business research firm Roots Analysis on 1 March 2023, driven by the development of new platforms and advanced technologies, as well as the unremitting efforts of relevant enterprises in the research and development of exosomes therapy, the exosomes therapy market is expected to grow at an annual growth rate of 41%. Due to the approval of many candidate drugs for exosome treatment in the coming years, the market is expected to reach billions of dollars.

Some innovative enterprises focusing on the research and application of exosomes have been sprung up in recent years: *Codiak BioSciences, Exosome Diagnostics, Exosome Sciences, Exovita Biosciences, BioRegenerative Sciences, ExoCoBio,* etc. Even larger corporations, such as IBM, have begun to march into the field of exosomes research.

7.4.1.1 ExoCoBio

ExoCoBio was established in January 2017 and is headquartered in Seoul, South Korea, developing cosmetic products based on stem cell-derived exosomes. The total financing amount of the company has reached approximately $56.3 million, with funds coming from top investors in South Korea, including Seven Tree Equity Partners, CSQUARED Global Asset Management, TS Investment Partners, K2 Investment, Invest, KDB Capital, Atinum Investment, GU Equity Partners, QUANTUM Ventures Korea, Platinum Technology Investment, and others.

ExoCoBio products include ASCE+, EXOMAGE, and CELLTWEET, used for skin care, regulating skin immune function, and scalp care. ASCE+ is a patented exosome regeneration technology from ExoCoBio, which utilizes 5 growth factors, 6 peptides, 19 amino acids, 4 different types of coenzymes, vitamins and minerals, glutathione, and various other active ingredients to achieve nutritional supply and stronger skin regeneration effects. It has a higher ability to deeply repair wounds, clean scars, and improve acne.

7.4.1.2 Direct Biologics

Direct Biologics is a regenerative biotechnology company with groundbreaking EV platform technology, headquartered in Austin, Texas, USA. The products under research include two series: ExoFlo and XoFlo. ExoFlo utilizes extracellular vesicles secreted by mesenchymal stem cells (MSCs) for tissue regeneration and repair. Direct Biologics is the first and only exosomes company to obtain FDA Phase III clinical approval, with the approved product being ExoFlo. It is an extracellular signaling product isolated from human bone marrow mesenchymal stem cells, containing growth factors and extracellular vesicles. Research has shown that extracellular vesicles derived from BM-MSC can downregulate human inflammation and upregulate tissue repair.

ExoFlo™ is currently being used to treat acute respiratory distress syndrome (ARDS) associated with COVID-19. On 21 April 2022, the FDA approved the company's use of ExoFlo for its Phase III clinical trial.

On 13 October 2022, the listed special purpose acquisition companies Good Works II Acquisition Corp. (Nasdaq: GWII) and Direct Biologics announced the signing of an agreement to become a listed company once all transaction conditions are met. The transaction is expected to be completed in the first half of 2023, and the merged company will be named Direct Biologics, Inc., which will be traded on the Nasdaq Capital Market.

7.4.1.3 Tianjin Exocrine Science and Technology

Tianjin Exocrine Science and Technology Incubation was established by Jiuzhoutong Group Tianjiu regenerative medicine in 2021. With exocrine technology as the core, Tianjin Exocrine Science and Technology Incubation focuses on serious medical and consumer medical business.

The company has built four major exosomes technology platforms: (i) a large-scale exosomes preparation platform, achieving the large-scale preparation of milk/293 cells/mesenchymal stem cell exosomes. With a single processing capacity of over 500 l, exosomes can be prepared on a large scale and with high purity, breaking the first bottleneck in the industrial application of exosomes. (ii) Pioneering exosomes surface modification loading platform with unique BioBIND® Technology can load a wide range of functional substances onto the surface of exosomes, and ensure the biological activity of exosomes and active substances. Through modular loading, the bottleneck of low drug loading and high safety risks for ADC drugs can be overcome. (iii) The TAD self-assembled gene drug loading technology platform with optimized design—optimized gene drug loading system—is used for the development of various gene drugs and mRNA vaccines. It has advantages such as low immunogenicity, large loaded gene fragments, prominent targeting effects, and oral administration. (iv) Lipid chimeric loading platform enables oral delivery of biological macromolecules and small molecules such as nucleic acids and peptides.

In terms of serious medical treatment, Tianjin Exosomes Technology has developed the "milk exosomes raw material," which is prepared by multilevel ultra-deep filtration chromatography technology and has a purity of over 99%; its platform technology can provide nano delivery system drug loading CRO and CDMO services, and jointly develop innovative drugs or second-class new drugs based on exosomes, which can load active substances such as antibodies, peptides, small molecules, nucleic acids, etc.

In terms of consumer healthcare, milk exosomes have been officially approved for inclusion in the International Cosmetic Ingredients (INCI) catalog, with the product name "Milk Exosomes" and INCI number "35079." The "Ran" series of exosomes peptide composite raw materials have also been developed. The company also independently develops medical device-grade liquid dressings and cosmetic-grade functional skincare products for medical beauty and skincare [3]- EXOSOMES LIMITED BANDAGE has obtained European Union (EU) and Conformité Europḗene (CE)

Certification, and can be sold in 27 EU member states and other countries that recognize EU CE certification.

7.4.1.4 TheraXyte

TheraXyte was established in 2019, focusing on the research and application of new nano drugs, especially innovative drugs related to exosomes. It is committed to the development and clinical application transformation of exosomes themselves, and their innovative technologies as drug delivery carriers. In addition, based on professional and comprehensive TAXY™—the exosomes research platform, the company also provides research CRO services and CDMO services in related fields to the outside world.

In July 2022, TheraXyte completed a pre-A round of financing of over ten million US dollars, led by Novartis Ventures.

TheraXyte has conducted comprehensive innovative research and development on extracellular drug delivery technology, and independently developed a programmed drug delivery system—TAXY. We have established an exosome production cell line ExoBoost with independent intellectual property rights, which can increase exosome production by 20 times compared to existing cell lines such as 293T and Expi293F. In addition, the company has developed the ExoPack exosomes drug loading system, which can efficiently and selectively load protein or nucleic acid drugs into the interior or surface of exosomes. Based on the different mechanisms by which exosomes enter cells and the different effects of receptor ligands, exosomes with different tissue targeting properties have been developed, which can target the brain, lungs, muscles, etc.

TheraXyte has nearly 1000 m^2 of GMP production and quality control area, and has accumulated rich development and production experience in the fields of plasmid construction, strain bank establishment and verification, cell bank establishment and verification, process development and optimization, QC testing and release, cell culture, and extracellular purification. Relying on its own extracellular drug research and development platform, production technology advantages, and a comprehensive multifunctional GMP workshop and GMP production system, the company now has the ability to prepare sample sizes from laboratory level (<2 l) to industrial level (200 l).

7.4.1.5 Exosome Diagnostics

Exosome Diagnostics was acquired by Bio Techne on 25 June 2018 for $250 million in cash and $325 million in milestone payments.

Exosome Diagnostics focuses on developing diagnostic technologies based on biological fluids, which can utilize extracellular vesicles such as blood, urine, and cerebrospinal fluid to recover RNA and DNA, and analyze specific genetic characteristics of diseases. The company has developed the ExoDx Lung (ALK) lung cancer diagnostic kit and ExoDx Prostate (IntelliScore) blood/urine detection kit. Patients can provide a blood or urine sample to analyze components such as RNA and protein from the obtained extracellular vesicles, thereby accurately diagnosing cancer. The company is able to develop new high-sensitivity diagnostic methods in various pathologies, including prostate cancer, bladder cancer, kidney cancer, breast cancer, glioblastoma, and other diseases, as well as cancers for which there is no diagnostic solution at present.

7.4.1.6 Codiak

Codiak was founded in 2015, and is a biotechnology company jointly promoted by venture capital firm ARCH Venture Partner and Flagship Pioneering. It was listed on the NASAQ stock market on 14 October 2020 under the stock code "CDAK." Codiak Biosciences raised $257.4 million in its A, B, and C rounds of financing, as well as three rounds of financing after its IPO. Due to the "capital winter" and the "suspension of two clinical phase II trial pipelines," the current market value only remains $18.05 million (see reason 5.2.7).

Codiak is committed to engineering exosomes, using exosomes to mediate drug delivery and precise treatment, and has the core patent platform engEx. It can produce exosomes of industrial grade and scale. Codiak discovered naturally occurring exosomes proteins—PTGFRN and BASP1—as scaffolds to guide targeted ligands and active molecules to the surface or lumen of exosomes. There are seven pipelines under research, with ExoIL-12, exoSTING, and exoASO-STAT6 in clinical phase I/II. However, the ExoIL-12 and exoSTING projects have been suspended, and the remaining products are in the preclinical research stage.

7.4.1.7 Evox

Evox Therapeutics was founded in April 2016 and conducted three rounds of financing in 2016, 2018, and 2021, with a total financing of $169.2 million, focusing on the treatment of rare diseases based on extracellular drug delivery.

DeliverEX developed by Evox™—The exosomes delivery technology platform can design and modify exosomes, and load proteins, RNA drugs, and small-molecule drugs into the modified exosomes to target tissues, especially for delivery to the brain and central nervous system. Evox also has its unique patented Scaffold protein TSPAN2. Compared with Codiak's PTGFRN Scaffold protein, Evox has advantages in drug loading and exosome specificity.

Using the DeliverEXTM platform, Evox is developing internal and partner exosomes therapeutic product pipelines for the treatment of rare, life-threatening diseases such as lysosomal storage syndrome, phenylketonuria, citrullinemia type I, etc. Evox collaborates with Takeda Pharmaceuticals to develop protein replacement therapy using exosomal vectors for the treatment of LSD and another rare disease, and cooperates with Lilly Pharmaceutical to develop RNAi and ASO drugs for the treatment of nervous system disease.

7.4.1.8 EVerZom

EVerZom was founded in September 2019 and is a start-up company headquartered in France, developing an extracellular vesicle production platform that meets GMP standards. As a CDMO company in the field of exosomes, EVErZom aims to overcome the challenge of large-scale production of exosomes through breakthrough patented production technology. The patent innovation of EVerZom lies in the application of surge stimulation to cells to trigger the release of a large amount of extracellular vesicles. By modifying a bioreactor to induce "turbulence stimulation," the production speed of extracellular vesicles is 10 times faster than traditional methods.

EVerZom provides customers with exosomes manufacturing, characterization, separation, and engineering services. It also simultaneously provides standards for adipose-derived mesenchymal stem cell extracellular vesicles and extracellular vesicles from mouse stem cells.

7.5 Issues and Challenges

7.5.1 Quality Control

Exosomes—a complex substance and structure that can only be produced by living cells—face a certain degree of inherent biological variation that leads to product heterogeneity, which is influenced by both intracellular biological processing and manufacturing processes. Culture conditions such as cell passage, cell density, and extracellular vesicle harvesting frequency can affect various aspects of product quality, including yield, extracellular vesicle composition, and extracellular vesicle's biological activity.

Detecting and evaluating changes in product quality during the production of extracellular vesicles is more challenging than in the production of proteins and antibodies. Due to the significant impact that small changes may have on product quality and activity, the stability of the exosomes manufacturing process is lower than that of the cell and antibody production processes; but exosomes manufacturing can still be built on these processes. The complexity of the size, composition, and structure of exosomes is unique; therefore, additional procedures are required to control the production of exosomes. These processes may include measures such as measuring size and concentration, removing pollutants, identifying functional markers, and evaluating drug loading to evaluate the therapeutic activity of each quantity of extracellular vesicles to ensure that the final product of the extracellular vesicles meets the previously determined key quality attributes.

7.5.2 Storage Stability

It is recommended to store exosomes at −80 °C, and their size and quantity remain unchanged at −80 °C. The potential problem is that they may cause some changes in biological activity, and the cost of storage and transportation is relatively high. Other studies have shown that storage at 4 °C can cause aggregation and damage to the extracellular structure.

Scientists have proposed alternative methods. Wu et al. proposed that storing exosomes in GelMA hydrogel can reduce exosomes aggregation and achieve the purpose of preserving exosomes. Another common preservation method is freeze-drying, where the freeze-dried extracellular vesicles can be stored at room temperature, and maintain key properties such as vesicle size and protein content, structural integrity, biological activity, and pharmacokinetics of drug molecules. The effect can be comparable to that of fresh extracellular vesicles.

7.5.3 Product Safety

Nanodrugs may cause some adverse immune reactions, such as allergic reactions, cytokine release syndrome, neutralization of biological activity, cross-reactions with endogenous protein counterparts, and nonacute immune reactions. Therefore, proving the safety and effectiveness of engineered exosomes is a challenge for researchers.

At present, product manufacturers are required to indicate all substances in the drug that can cause pharmacological, immune, or metabolic effects, and the inactive ingredients used in the final formula (referred to as excipients) should also be declared. The International Association of Extracellular Vesicles states that if the entire therapeutic effect can be attributed to the drug loaded on the exosomes, rather than the exosomes in the drug-loaded exosomes system, the exosomes can be considered excipients. In this case, only the safety description of the exosomes is needed, without describing their mode of action or mechanism. If both the safety and quality standards fully meet the requirements, and provide reliable methods and mechanisms of action, human clinical trials can be allowed.

To convert allogeneic exosomes and heterologous exosomes into safe and effective clinical treatment products, it is necessary to strictly control the selection of parental cells, immunology, evaluation of carcinogenic effects, and risk of virus contamination through continuous monitoring. The selection of monitoring methods, especially sensitivity, is one of the key challenges in determining the clinical conversion of extracellular drug carriers. At present, regulatory agencies have not yet issued guidance on how to test the safety and effectiveness of exosomes, so it is necessary to improve existing testing methods, or find new methods to prove the safety and effectiveness of engineered exosomes.

7.5.4 Drug-Forming Properties of Engineered Exosomes

Natural exosomes have been launched and sold in the field of beauty and skincare, and the pharmaceutical field has entered clinical phase III, with high commercial value and certainty of drug formation.

The engineering exosomes mainly leverage their advantages in delivering active ingredients and carrying "which active ingredient" directly affects the success or failure of drugs. The suspension of Codiak's clinical pipeline has a significant impact on industry confidence. Codiak, as a benchmark enterprise for engineering exosomes for drug delivery, has always been highly concerned and has solved problems such as exosomes skeleton protein screening, endogenous loading technology, and large-scale production. The two drug pipelines that Codiak first entered clinical practice in 2022 are ExoIL-12 and exoSTING (Please refer to Section 2.2.1 for protein loading), and the treatment method is "intratumor injection." Phase I clinical trial results were released in mid-2022: ExoIL-12. The safety is good, and the primary endpoint was achieved in the Phase I clinical trial, and both the patients were treated and showed partial remission. Plasma pharmacokinetic (PK) measurements of subjects that received exoSTING showed no systemic exposure to the agonist. Further, analyses of available plasma biomarkers indicated a lack of systemic inflammatory cytokines detectable in the blood after exoSTING administration. All the results supported the product's promotion to clinical phase II, but Codiak announced in August 2022 that it would suspend the promotion of phase II clinical trials for both products, causing a sharp drop in stock prices. On 27 March 2023, it declared bankruptcy protection, causing a great uproar in the industry and leading to a major discussion on whether engineering exosomes as nanodelivery drug delivery systems can be successfully commercialized.

References

1 Johnstone, R.M., Adam, M., Hammond, J.R. et al. (1987). Vesicle formation during reticulocyte maturation. Association of plasma membrane activities with released vesicles (exosomes). *J. Biolumin. Chemilumin.* 262 (19): 9412–9420.

2 Caby, M.P., Lankar, D., Vincendeau-Scherrer, C. et al. (2005). Exosomal-like vesicles are present in human blood plasma. *Int. Immunol.* 17 (7): 879–887.

3 Pisitkun, T., Shen, R.F., and Knepper, M.A. (2004). Identification and proteomic profiling of exosomes in human urine. *Proc. Natl. Acad. Sci. U.S.A.* 101 (36): 13368–13373.

4 Street, J.M., Barran, P.E., Mackay, C.L. et al. (2012). Identification and proteomic profiling of exosomes in human cerebrospinal fluid. *J. Transl. Med.* 10: 5.

5 Palanisamy, V., Sharma, S., Deshpande, A. et al. (2010). Nanostructural and transcriptomic analyses of human saliva derived exosomes. *PLoS One* 5 (1): e8577.

6 Bard, M.P., Hegmans, J.P., Hemmes, A. et al. (2004). Proteomic analysis of exosomes isolated from human malignant pleural effusions. *Am. J. Respir. Cell Mol. Biol.* 31 (1): 114–121.

7 Runz, S., Keller, S., Rupp, C. et al. (2007). Malignant ascites-derived exosomes of ovarian carcinoma patients contain CD24 and EpCAM. *Gynecol. Oncol.* 107 (3): 563–571.

8 Keller, S., Rupp, C., Stoeck, A. et al. (2007). CD24 is a marker of exosomes secreted into urine and amniotic fluid. *Kidney Int.* 72 (9): 1095–1102.

9 Admyre, C., Johansson, S.M., Qazi, K.R. et al. (2007). Exosomes with immune modulatory features are present in human breast milk. *J. Immunol.* 179 (3): 1969–1978.

10 Kulshreshtha, A., Ahmad, T., Agrawal, A., and Ghosh, B. (2013). Proinflammatory role of epithelial cell-derived exosomes in allergic airway inflammation. *J. Allergy Clin. Immunol.* 131 (4): 1194–1203, 203.e1-14.

11 Yáñez-Mó, M., Siljander, P.R., Andreu, Z. et al. (2015). Biological properties of extracellular vesicles and their physiological functions. *J. Extracell. Vesicles* 4: 27066.

12 Yu, D., Li, Y., Wang, M. et al. (2022). Exosomes as a new frontier of cancer liquid biopsy. *Mol. Cancer* 21 (1): 56.

13 Kalluri, R. and LeBleu, V.S. (2020). The biology, function, and biomedical applications of exosomes. *Science* 367 (6478): eaau6977.

14 Jansen, F. and Li, Q. (2017). Exosomes as Diagnostic Biomarkers in Cardiovascular Diseases. *Adv. Exp. Med. Biol.* 998: 61–70.

15 Wong, L.L., Armugam, A., Sepramaniam, S. et al. (2015). Circulating microRNAs in heart failure with reduced and preserved left ventricular ejection fraction. *Eur. J. Heart Failure* 17 (4): 393–404.

16 Kanninen, K.M., Bister, N., Koistinaho, J., and Malm, T. (2016). Exosomes as new diagnostic tools in CNS diseases. *Biochim. Biophys. Acta* 1862 (3): 403–410.

17 Fitts, C.A., Ji, N., Li, Y., and Tan, C. (2019). Exploiting exosomes in cancer liquid biopsies and drug delivery. *Adv. Healthc. Mater.* 8 (6): e1801268.

18 Kalluri, R. (2016). The biology and function of exosomes in cancer. *J. Clin. Invest.* 126 (4): 1208–1215.

19 Masyuk, A.I., Masyuk, T.V., and Larusso, N.F. (2013). Exosomes in the pathogenesis, diagnostics and therapeutics of liver diseases. *J. Hepatol.* 59 (3): 621–625.

20 Zhang, W., Zhou, X., Zhang, H. et al. (2016). Extracellular vesicles in diagnosis and therapy of kidney diseases. *Am. J. Physiol. Renal Physiol.* 311 (5): F844–f51.

21 Alipoor, S.D., Mortaz, E., Garssen, J. et al. (2016). Exosomes and exosomal miRNA in respiratory diseases. *Mediators Inflammation* 2016: 5628404.

22 Braicu, C., Tomuleasa, C., Monroig, P. et al. (2015). Exosomes as divine messengers: are they the Hermes of modern molecular oncology? *Cell Death Differ.* 22 (1): 34–45.

23 Sun, Y., Gao, Y., Sun, J. et al. (2016). Expression profile analysis based on DNA microarray for patients undergoing off-pump coronary artery bypass surgery. *Exp. Ther. Med.* 11 (3): 864–872.

24 Xu, Y., Zhu, W., Sun, Y. et al. (2015). Functional network analysis reveals versatile microRNAs in human heart. *Cell. Phys. Biochem.* 36 (4): 1628–1643.

25 Crouser, E.D., Hamzeh, N.Y., Maier, L.A. et al. (2017). Exosomal microRNA for detection of cardiac sarcoidosis. *Am. J. Respir. Crit. Care Med.* 196 (7): 931–934.

26 Selmaj, I., Cichalewska, M., Namiecinska, M. et al. (2017). Global exosome transcriptome profiling reveals biomarkers for multiple sclerosis. *Ann. Neurol.* 81 (5): 703–717.

27 Schwarzenbach, H. (2015). The clinical relevance of circulating, exosomal miRNAs as biomarkers for cancer. *Expert Rev. Mol. Diagn.* 15 (9): 1159–1169.

28 Yang, H., Fu, H., Xu, W., and Zhang, X. (2016). Exosomal non-coding RNAs: a promising cancer biomarker. *Clin. Chem. Lab. Med.* 54 (12): 1871–1879.

29 Shyamala, K., Girish, H.C., and Murgod, S. (2014). Risk of tumor cell seeding through biopsy and aspiration cytology. *J. Int. Soc. Prev. Commun. Dent.* 4 (1): 5–11.

30 Cai, J., Han, Y., Ren, H. et al. (2013). Extracellular vesicle-mediated transfer of donor genomic DNA to recipient cells is a novel mechanism for genetic influence between cells. *J. Mol. Cell. Biol.* 5 (4): 227–238.

31 Kahlert, C., Melo, S.A., Protopopov, A. et al. (2014). Identification of double-stranded genomic DNA spanning all chromosomes with mutated KRAS and p53 DNA in the serum exosomes of patients with pancreatic cancer. *J. Biolumin. Chemilumin.* 289 (7): 3869–3875.

32 Kalluri, R. and LeBleu, V.S. (2016). Discovery of double-stranded genomic DNA in circulating exosomes. *Cold Spring Harbor Symp. Quant. Biol.* 81: 275–280.

33 Thakur, B.K., Zhang, H., Becker, A. et al. (2014). Double-stranded DNA in exosomes: a novel biomarker in cancer detection. *Cell Res.* 24 (6): 766–769.

34 Salehi, M. and Sharifi, M. (2018). Exosomal miRNAs as novel cancer biomarkers: challenges and opportunities. *J. Cell. Physiol.* 233 (9): 6370–6380.

35 Valadi, H., Ekström, K., Bossios, A. et al. (2007). Exosome-mediated transfer of mRNAs and microRNAs is a novel mechanism of genetic exchange between cells. *Nat. Cell Biol.* 9 (6): 654–659.

36 Taylor, D.D. and Gercel-Taylor, C. (2008). MicroRNA signatures of tumor-derived exosomes as diagnostic biomarkers of ovarian cancer. *Gynecol. Oncol.* 110 (1): 13–21.

37 Rabinowits, G., Gerçel-Taylor, C., Day, J.M. et al. (2009). Exosomal microRNA: a diagnostic marker for lung cancer. *Clin. Lung Cancer* 10 (1): 42–46.

38 Cazzoli, R., Buttitta, F., Di Nicola, M. et al. (2013). microRNAs derived from circulating exosomes as noninvasive biomarkers for screening and diagnosing lung cancer. *J. Thorac. Oncol.* 8 (9): 1156–1162.

39 Armstrong, D.A., Green, B.B., Seigne, J.D. et al. (2015). MicroRNA molecular profiling from matched tumor and bio-fluids in bladder cancer. *Mol. Cancer* 14: 194.

40 Samsonov, R., Shtam, T., Burdakov, V. et al. (2016). Lectin-induced agglutination method of urinary exosomes isolation followed by mi-RNA analysis: application for prostate cancer diagnostic. *Prostate* 76 (1): 68–79.

41 Liu, J., Ren, L., Li, S. et al. (2021). The biology, function, and applications of exosomes in cancer. *Acta Pharm. Sin. B* 11 (9): 2783–2797.

42 Melo, S.A., Luecke, L.B., Kahlert, C. et al. (2015). Glypican-1 identifies cancer exosomes and detects early pancreatic cancer. *Nature* 523 (7559): 177–182.

43 Gao, H.J., Guo, W., Xie, X.H. et al. (2022). Expression and diagnostic value of plasma exosomal hsa_circ_0064910 in gastric cancer. *Chi. J. Lab. Med.* 45 (3): 246–251.

44 Wu, H., Zhou, J., Mei, S. et al. (2017). Circulating exosomal microRNA-96 promotes cell proliferation, migration and drug resistance by targeting LMO7. *J. Cell. Mol. Med.* 21 (6): 1228–1236.

45 Chen, R., Xu, X., Qian, Z. et al. (2019). The biological functions and clinical applications of exosomes in lung cancer. *Cell. Mol. Life Sci.* 76 (23): 4613–4633.

46 Jin, X., Chen, Y., Chen, H. et al. (2017). Evaluation of tumor-derived exosomal miRNA as potential diagnostic biomarkers for early-stage non-small cell lung cancer using next-generation sequencing. *Clin. Cancer Res.* 23 (17): 5311–5319.

47 Xu, Z.H., Miao, Z.W., Jiang, Q.Z. et al. (2019). Brain microvascular endothelial cell exosome-mediated S100A16 up-regulation confers small-cell lung cancer cell survival in brain. *FASEB J.* 33 (2): 1742–1757.

48 Lebanony, D., Benjamin, H., Gilad, S. et al. (2009). Diagnostic assay based on hsa-miR-205 expression distinguishes squamous from nonsquamous non-small-cell lung carcinoma. *J. Clin. Oncol.* 27 (12): 2030–2037.

49 Gao, W., Yu, Y., Cao, H. et al. (2010). Deregulated expression of miR-21, miR-143 and miR-181a in non small cell lung cancer is related to clinicopathologic characteristics or patient prognosis. *Biomed. Pharm.* 64 (6): 399–408.

50 Liu, C., Xu, X., Li, B. et al. (2018). Single-exosome-counting immunoassays for cancer diagnostics. *Nano Lett.* 18 (7): 4226–4232.

51 Sohn, W., Kim, J., Kang, S.H. et al. (2015). Serum exosomal microRNAs as novel biomarkers for hepatocellular carcinoma. *Exp. Mol. Med.* 47 (9): e184.

52 Tang, J., Jiang, R., Deng, L. et al. (2015). Circulation long non-coding RNAs act as biomarkers for predicting tumorigenesis and metastasis in hepatocellular carcinoma. *Oncotarget* 6 (6): 4505–4515.

53 Ma, X., Yuan, T., Yang, C. et al. (2017). X-inactive-specific transcript of peripheral blood cells is regulated by exosomal Jpx and acts as a biomarker for female patients with hepatocellular carcinoma. *Ther. Adv. Med. Oncol.* 9 (11): 665–677.

54 Yu, L.X., Zhang, B.L., Yang, Y. et al. (2019). Exosomal microRNAs as potential biomarkers for cancer cell migration and prognosis in hepatocellular carcinoma patient-derived cell models. *Oncol. Rep.* 41 (1): 257–269.

55 Sun, L., Su, Y., Liu, X. et al. (2018). Serum and exosome long non coding RNAs as potential biomarkers for hepatocellular carcinoma. *J. Cancer* 9 (15): 2631–2639.

56 Wang, G., Liu, W., Zou, Y. et al. (2019). Three isoforms of exosomal circPTGR1 promote hepatocellular carcinoma metastasis via the miR449a-MET pathway. *EBioMedicine* 40: 432–445.

57 Matsuo, K., Hasegawa, K., Yoshino, K. et al. (2015). Venous thromboembolism, interleukin-6 and survival outcomes in patients with advanced ovarian clear cell carcinoma. *Eur. J. Cancer* 51 (14): 1978–1988.

58 Zhang, W., Ou, X., and Wu, X. (2019). Proteomics profiling of plasma exosomes in epithelial ovarian cancer: a potential role in the coagulation cascade, diagnosis and prognosis. *Int. J. Oncol.* 54 (5): 1719–1733.

59 Boussios, S., Rassy, E., Samartzis, E. et al. (2021). Melanoma of unknown primary: new perspectives for an old story. *Crit. Rev. Oncol./Hematol.* 158: 103208.

60 Zebrowska, A., Widlak, P., Whiteside, T., and Pietrowska, M. (2020). Signaling of tumor-derived sEV impacts melanoma progression. *Int. J. Mol. Sci.* 21 (14): 5066.

61 Hood, J.L. (2019). Natural melanoma-derived extracellular vesicles. *Semin. Cancer Biol.* 59: 251–265.

62 Bollard, S.M., Casalou, C., Goh, C.Y. et al. (2020). Circulating melanoma-derived extracellular vesicles: impact on melanoma diagnosis, progression monitoring, and treatment response. *Pharmaceuticals* 13 (12): 475.

63 Gowda, R., Robertson, B.M., Iyer, S. et al. (2020). The role of exosomes in metastasis and progression of melanoma. *Cancer Treat. Rev.* 85: 101975.

64 Isola, A.L., Eddy, K., and Chen, S. (2016). Biology, therapy and implications of tumor exosomes in the progression of melanoma. *Cancers* 8 (12): 110.

65 Zhang, Z., Liu, X., Yang, X. et al. (2023). Identification of faecal extracellular vesicles as novel biomarkers for the non-invasive diagnosis and prognosis of colorectal cancer. *J. Extracell. Vesicles* 12 (1): e12300.
66 Sung, H., Ferlay, J., Siegel, R.L. et al. (2021). Global cancer statistics 2020: GLOBOCAN estimates of incidence and mortality worldwide for 36 cancers in 185 countries. *Cancer J. Clin.* 71 (3): 209–249.
67 Wang, H., Jiang, D., Li, W. et al. (2019). Evaluation of serum extracellular vesicles as noninvasive diagnostic markers of glioma. *Theranostics* 9 (18): 5347–5358.
68 Coons, S.W., Johnson, P.C., and Pearl, D.K. (1997). The prognostic significance of Ki-67 labeling indices for oligodendrogliomas. *Neurosurgery* 41 (4): 878–884; discussion 84-5.
69 Scholzen, T. and Gerdes, J. (2000). The Ki-67 protein: from the known and the unknown. *J. Cell. Physiol.* 182 (3): 311–322.
70 Yerushalmi, R., Woods, R., Ravdin, P.M. et al. (2010). Ki67 in breast cancer: prognostic and predictive potential. *Lancet Oncol.* 11 (2): 174–183.
71 Castellanos-Rizaldos, E., Grimm, D.G., Tadigotla, V. et al. (2018). Exosome-based detection of EGFR T790M in plasma from non-small cell lung cancer patients. *Clin. Cancer Res.* 24 (12): 2944–2950.
72 Zhou, B., Xu, K., Zheng, X. et al. (2020). Application of exosomes as liquid biopsy in clinical diagnosis. *Signal Transduct. Target Ther.* 5 (1): 144.
73 McKiernan, J., Donovan, M.J., O'Neill, V. et al. (2016). A novel urine exosome gene expression assay to predict high-grade prostate cancer at initial biopsy. *JAMA Oncol.* 2 (7): 882–889.
74 McKiernan, J., Donovan, M.J., Margolis, E. et al. (2018). A prospective adaptive utility trial to validate performance of a novel urine exosome gene expression assay to predict high-grade prostate cancer in patients with prostate-specific antigen 2-10ng/ml at initial biopsy. *Eur. Urol.* 74 (6): 731–738.
75 Margolis, E., Brown, G., Partin, A. et al. (2022). Predicting high-grade prostate cancer at initial biopsy: clinical performance of the ExoDx (EPI) Prostate Intelliscore test in three independent prospective studies. *Prostate Cancer Prostatic Dis.* 25 (2): 296–301.
76 Fu, W., Lei, C., Liu, S. et al. (2019). CAR exosomes derived from effector CAR-T cells have potent antitumour effects and low toxicity. *Nat. Commun.* 10 (1): 4355.
77 Hansen, A.L. (ed.) (2013). *Cancer Immunology and Immunotherapy*. Keystone Symposia.
78 Krug, A.K., Enderle, D., Karlovich, C. et al. (2018). Improved EGFR mutation detection using combined exosomal RNA and circulating tumor DNA in NSCLC patient plasma. *Ann. Oncol.* 29 (10): 2143.
79 Agrawal, A.K., Aqil, F., Jeyabalan, J. et al. (2017). Milk-derived exosomes for oral delivery of paclitaxel. *Nanomedicine* 13 (5): 1627–1636.
80 Allenson, K., Castillo, J., San Lucas, F.A. et al. (2017). High prevalence of mutant KRAS in circulating exosome-derived DNA from early-stage pancreatic cancer patients. *Ann. Oncol.* 28: 741–747.
81 Kimura, H., Yamamoto, H., Harada, T. et al. (2019). CKAP4, a DKK1 receptor, is a biomarker in exosomes derived from pancreatic cancer and a molecular target for therapy. *Clin. Cancer. Res.* 25: 1936–1947.

82 Dong, L., Lin, W., Qi, P. et al. (2016). Circulating long RNAs in serum extracellular vesicles: their characterization and potential application as biomarkers for diagnosis of colorectal cancer. *Cancer Epidemiol. Biomarkers Prev.* 25: 1158–1166.

83 Zhao, L., Yu, J., Wang, J. et al. (2017). Isolation and Identification of miRNAs in exosomes derived from serum of colon cancer patients. *J. Cancer* 8: 1145–1152.

84 Li, J., Chen, Y., Guo, X. et al. (2017). GPC1 exosome and its regulatory miRNAs are specific markers for the detection and target therapy of colorectal cancer. *J. Cell. Mol. Med.* 21 (5): 838–847.

85 Yokoyama, S., Takeuchi, A., Yamaguchi, S. et al. (2017). Clinical implications of carcinoembryonic antigen distribution in serum exosomal fraction-measurement by ELISA. *PLoS. One* 12: e0183337.

86 Sun, B., Li, Y., Zhou, Y. et al. (2019). Circulating exosomal CPNE3 as a diagnostic and prognostic biomarker for colorectal cancer. *J. Cell. Physiol.* 234: 1416–1425.

87 Rodriguez, M., Silva, J., López-Alfonso, A. et al. (2014). Different exosome cargo from plasma/bronchoalveolar lavage in non-small-cell lung cancer. *Genes. Chromosomes Cancer* 53: 713–724.

88 Sandfeld-Paulsen, B., Jakobsen, K.R., Bak, R. et al. (2016). Exosomal Proteins as Diagnostic Biomarkers in Lung Cancer. *J. Thorac. Oncol* 11: 1701–1710.

89 Niu, L., Song, X., Wang, N. et al. (2019). Tumor-derived exosomal proteins as diagnostic biomarkers in non-small cell lung cancer. *Cancer Sci.* 110 (1): 433–442.

90 Hannafon, B.N., Trigoso, Y.D., Calloway, C.L. et al. (2016). Plasma exosome microRNAs are indicative of breast cancer. *Breast Cancer Res.* 18 (1): 90.

91 Zhou, W., Fong, M.Y., Min, Y. et al. (2014). Cancer-secreted miR-105 destroys vascular endothelial barriers to promote metastasis. *Cancer Cell.* 25: 501–515.

92 Wang X, Zhong W, Bu J et al. (2019) May; Exosomal protein CD82 as a diagnostic biomarker for precision medicine for breast cancer. *Mol. Carcinog.* 58 (5): 674–685.

93 Chen, I.H., Xue, L., Hsu, C.C. et–al. (2017). Phosphoproteins in extracellular vesicles as candidate markers for breast cancer. *Proc Natl Acad Sci U S A* 114 (12): 3175–3180.

94 Raimondo, F., Morosi, L., Corbetta, S. et al. (2013). Differential protein profiling of renal cell carcinoma urinary exosomes. *Mol Biosyst* 9: 1220–1233.

95 Matsuzaki, K., Fujita, K., Jingushi, K. et al. (2017). MiR-21-5p in urinary extracellular vesicles is a novel biomarker of urothelial carcinoma. *Oncotarget* 8: 24668–24678.

96 Arbelaiz, A., Azkargorta, M., Krawczyk, M. et al. (2017). Serum extracellular vesicles contain protein biomarkers for primary sclerosing cholangitis and cholangiocarcinoma. *Hepatology* 66: 1125–1143.

8

Conclusion and Future Perspectives

Xing-Jie Liang[1,2], *Ke Cheng*[3], *and Zhenhua Li*[4,5]

[1] *National Center for Nanoscience and Technology of China, Chinese Academy of Sciences (CAS) Center for Excellence in Nanoscience, CAS Key Laboratory for Biomedical Effects of Nanomaterials and Nanosafety, 11, Beiyitiao Alley, Beijing, 100190, China*
[2] *University of Chinese Academy of Sciences, CAS Key Laboratory for Biomedical Effects of Nanomaterials and Nanosafety, 11, Beiyitiao Alley, Beijing, 100049, China*
[3] *Department of Biomedical Engineering, Columbia University, New York 10032, USA*
[4] *The Tenth Affiliated Hospital of Southern Medical University, 78 Wanjiang Avenue, Dongguan, Guangdong 523059, China*
[5] *Guangdong Provincial Key Laboratory of Shock and Microcirculation, Shatai South Avenue, Guangzhou, Guangdong 510080, China*

8.1 Summary and Conclusions

EVs are a type of naturally occurring cell secretion that has gained increasing attention in the recent years due to their diverse biological functions and potential biomedical applications. EVs, including exosomes, microvesicles, and ectosomes, can be found in various biological fluids, such as blood, urine, and saliva, and play crucial roles in physiological and pathological processes, such as cell signaling, apoptosis, and immune regulation. EVs are a promising tool for biomedical applications. They are a natural secretion of cells that play important roles in different physiological and pathological processes. This book aimed to provide an introduction into this rapidly evolving and dynamic field of research. In the eight chapters that form the core of this book, more than 30 authors have documented the current state of the field of EVs in biomedical applications. In Chapter 1, we have an overview about EVs and their various biomedical applications. In Chapter 2, we focus on the biogenesis, identification, and their compositions with different biofunctions of EVs. Chapter 3 is mainly for discussion of EVs from stem cells, immune cells, tumor cells, and other cells (plants, microorganisms) for their biofunctions such as anti-inflammation, promoting tissue repair and regeneration, and neuroprotection. In Chapter 4, we introduce the EVs derived from different source in the treatment of various diseases by taking advantage of their specific biofunctions of different EVs. Considering EVs themselves may lead to off-target

Biomedical Applications of Extracellular Vesicles, First Edition.
Edited by Zhenhua Li, Xing-Jie Liang, and Ke Cheng.

effect and low retention, we have focused on engineered EVs for biomedical application in Chapter 5. In Chapter 6, we discuss the current technology for EVs production, isolation, and quality control due to their limitations in the manufacturing. Chapter 7 includes the recent progress of EVs on clinical trials, and their future prospects and limitations of clinical applications.

In the concluding chapter, we will first summarize the main findings of each part of the book. Next, we will discuss some general trends and developments, followed by a discussion of key challenges for the future research in biomedical applications of EVs.

8.2 General Trends and Developments

In the previous section we have already introduced many trends and developments within specific research areas of EVs in biomedical applications. In this section, we will further discuss three more general, overarching trends and developments, as they emerge from the whole of this book.

8.2.1 EVs in Drug Delivery

EVs possess natural low immunogenicity, low toxicity, and high biostability, making them ideal drug carriers for various molecules such as proteins, nucleic acids, and drugs. For example, EVs are natural carriers of cargo molecules, and their lipid bilayer structure enables them to cross biological barriers such as the blood–brain barrier. This makes them an attractive option for delivering therapeutic molecules to target tissues and organs. In personalized medicine, EVs can be isolated from patient-specific cells and are used for personalized medicine. This approach can improve treatment efficacy and reduce side effects. In targeted drug delivery, EVs can be engineered to express specific surface proteins or peptides that target specific cells or tissues. This allows for targeted drug delivery and reduces off-target effects. For example, EVs can be engineered to express a tumor-targeting peptide, improving their specificity for cancer cells. In addition, EVs can be used for sustained drug release. EVs can be loaded with drugs and released over an extended period, providing sustained therapeutic effects. EVs from immune cells possess immunomodulatory properties. EVs can modulate the immune response, making them attractive for immunotherapy applications. For example, EVs derived from antigen-presenting cells can be used to induce immune responses against cancer cells.

There are many mRNA-based drug delivery systems that have developed since mRNA vaccines have shown high therapeutic potential in COVID-19 and other critical diseases. We would like to have a brief discussion about EVs for gene delivery to highlight their key role in drug delivery. Recently, there has been growing interest in using EVs for gene delivery, as they offer several advantages over traditional gene delivery methods, such as viral vectors and liposomes. One of the main advantages of using EVs for gene delivery is their ability to naturally transport cargo (such as DNA or RNA) across biological barriers, such as the blood–brain barrier. Additionally,

EVs are nonimmunogenic, meaning they do not elicit an immune response in the recipient, reducing the risk of adverse reactions. Furthermore, EVs for gene delivery is their natural ability to penetrate the cell membrane and deliver their cargo to the cytoplasm of target cells. This makes EVs an attractive alternative to viral vectors, which are associated with safety concerns and have limited cargo capacity. Research in this area has focused on understanding the mechanisms by which EVs interact with target cells and deliver their cargo, as well as optimizing methods for producing and isolating EVs for gene delivery. Various sources of EVs have been explored for gene delivery, including mesenchymal stem cells, immune cells, and cancer cells. They have been shown to be capable of delivering various types of nucleic acids, including small interfering RNA (siRNA), microRNA (miRNA), and messenger RNA (mRNA). There are several strategies for loading nucleic acids into EVs. One approach is to engineer the parental cells to overexpress the desired nucleic acid, which will then be incorporated into the EVs during their biogenesis. Another approach is to directly load the nucleic acid into isolated EVs using electroporation, sonication, or other physical methods. Additionally, modifications to the EV membrane, such as the addition of targeting peptides or antibodies, can be used to enhance the specificity and efficiency of gene delivery to specific cell types.

Overall, the use of EVs as gene delivery vehicle is a promising area of research that has the potential to overcome some of the limitations of existing gene delivery methods. However, further studies are needed to optimize the loading and targeting of EVs, and to evaluate their safety and efficacy in preclinical and clinical settings.

8.2.2 Engineered EVs in Biomedical Applications

EVs are nanoscale membrane-bound particles that play important roles in intercellular communication. Recently, there has been growing interest in the potential use of engineered EVs as therapeutics in biomedical applications due to their ability to deliver biologically active molecules to target cells and tissues.

In the following, here are some general trends and developments in the field of engineered EVs in biomedical engineering. The first is the engineering strategies of EVs. Up to now, there are various strategies that have been developed to engineer EVs, including loading with therapeutic cargo such as proteins, RNA, and DNA, modifying the surface of EVs to enhance targeting and uptake, and creating synthetic EVs by using biomimetic materials. Engineered EVs have been widely used in the area of targeted delivery. Engineered EVs can be targeted to specific cells or tissues by modifying their surface with targeting moieties such as antibodies, peptides, or aptamers. This allows for more precise delivery of therapeutics and reduces off-target effects. In addition, engineered EVs have more therapeutic applications. Engineered EVs have shown promise in a variety of therapeutic applications, including cancer therapy, regenerative medicine, and drug delivery. For example, engineered EVs loaded with siRNA or miRNA have been shown to inhibit tumor growth and metastasis. As for their manufacturing, the manufacturing of engineered EVs is a challenge due to their small size and complex composition.

However, the recent advances in microfluidics and extrusion technologies have enabled scalable production of EVs for clinical applications. The last one is the clinical translation. The translation of engineered EVs into clinical applications is still in its early stages. However, several companies are developing EV-based therapeutics for various diseases, and preclinical studies have shown promising results.

Overall, engineered EVs have great potential as a new class of therapeutics in biomedical engineering. Ongoing research and development in this field are likely to lead to further improvements in the design and manufacturing of EVs for clinical applications.

8.2.3 EVs for Clinical Applications When Comparing with Liposomes

When we talk about EVs for clinical applications, we would like to compare them with liposomes, as they have the similar size and structure. Liposomes have been widely used in clinical applications. Liposomes are versatile drug delivery systems that can encapsulate a variety of therapeutic molecules, including drugs, peptides, proteins, and nucleic acids. Liposomes can improve the pharmacokinetics and bioavailability of these molecules by protecting them from degradation and clearance in the bloodstream, and by targeting them to specific cells and tissues. Some examples of liposome-based therapeutics that have been approved for clinical use are Doxil, Ambisome, DepoDur and the recent mRNA-loaded COVID-19 vaccines. EVs are also nanoscale structures, and there are similarities between the two systems in clinical applications. However, there are some differences between these two approaches that are worth considering.

Natural versus synthetic: EVs are natural vesicles released by cells, whereas liposomes are synthetic vesicles made from lipids. This means that EVs have a more complex composition than liposomes, containing a variety of bioactive molecules such as proteins, nucleic acids, and lipids. Liposomes can be engineered to contain specific molecules, but the range of molecules that can be incorporated into liposomes is more limited than that of EVs.

Size and stability: EVs are typically smaller than liposomes and have greater stability *in vivo*. EVs are thought to be more resistant to degradation by enzymes and clearance by the immune system than liposomes. This may make EVs a more attractive option for drug delivery applications that require long circulation times.

Immunogenicity: EVs are less immunogenic than liposomes, as they are derived from cells and can be recognized as self by the immune system. Liposomes, on the other hand, can be recognized as foreign and may elicit an immune response.

Targeting: Both EVs and liposomes can be engineered to target specific cells or tissues. However, EVs have been shown to have a greater ability to target specific cells than liposomes. This is thought to be attributed to the presence of surface proteins and other molecules on EVs that allow them to interact with target cells.

Manufacturing: The production of EVs for clinical use can be challenging, as it often involves isolation from cell cultures or biological fluids. In contrast, liposomes can be synthesized in a controlled manner using a range of techniques, making their production more scalable and consistent.

In summary, while both EVs and liposomes have potential as drug delivery systems for clinical applications, they have different properties that may make them more suitable for specific applications. EVs have natural biocompatibility and targeting properties, while liposomes can be more stable and scalable. The choice between EVs and liposomes for a specific application will depend on the specific requirements of the drug delivery system. However, ongoing research and development are aimed at overcoming the challenges associated with scalability, reproducibility, and standardization of EV production, as well as the need to optimize loading efficiency and cargo stability.

8.3 Challenges for Future Research

Although the field of EVs in biomedical applications has steadily advanced over the past years to the stage where it has become an active contributor to a more sustainable society, many challenges for future research still exist. In this last section of the book, we discuss three major challenges that apply to the field as a whole. These challenges address the needs for further integration, further development of theories and models, and further engagement of environmental psychologists with environmental problems.

EVs have emerged as a promising platform for biomedical applications, especially in drug delivery, diagnostics, and regenerative medicine. However, there are still several challenges that need to be addressed before EV-based therapies can be widely adopted in the clinic. In the last section, we will discuss some of the key challenges for future research in the field of EVs for biomedical applications.

8.3.1 Standardization and Quality Control

One of the main challenges in the field of EVs is the lack of standardization and quality control. There is currently no consensus on the methods for EV isolation, characterization, and quantification, which can lead to variability in experimental results and difficulties in comparing data across studies. Standardization of these methods is critical to ensure the reproducibility and reliability of EV-based therapies.

Several initiatives have been launched to address this issue, including the International Society for Extracellular Vesicles (ISEVs) and the Minimal Information for Studies of Extracellular Vesicles (MISEVs) guidelines. These guidelines provide recommendations for EV isolation, characterization, and reporting of experimental data. However, it is important that these guidelines are widely adopted by the research community to ensure consistent and high-quality research.

8.3.2 Scalability and Manufacturing

Another challenge for EV-based therapies is the scalability and manufacturing of EVs. Current methods for EV isolation, such as ultracentrifugation and size exclusion chromatography, are time consuming and inefficient, and may not be suitable for large-scale production. There is a need for more efficient and scalable methods for EV isolation and purification.

In addition, the production of EVs from a consistent and reliable source is also a challenge. EVs derived from different cell types or under different conditions may have different properties and therapeutic potential. Furthermore, the yield of EVs from different cell types can vary widely, which can limit their clinical applicability. The development of standardized protocols for EV production and purification from a consistent and reliable source is necessary to enable the large-scale production of EVs for clinical use.

8.3.3 Targeting and Biodistribution

One of the key advantages of EVs as a drug delivery platform is their ability to target specific cells and tissues. However, the targeting efficiency of EVs can be limited by their biodistribution and clearance from the bloodstream. EVs can be rapidly cleared by the liver and spleen, which can reduce their therapeutic efficacy.

There is a need for more efficient and specific targeting strategies for EVs. Several approaches have been proposed, including the modification of EVs with targeting ligands or engineering EVs to express specific surface proteins. However, these approaches need to be further optimized to achieve efficient and selective targeting of EVs to specific cells and tissues.

8.3.4 Safety and Toxicity

Demonstration of the safety and efficacy of EV-based therapies is another important challenge in EV research. While there is a growing evidence for the potential therapeutic benefits of EVs, there is still a lack of data on their long-term safety and efficacy in humans.

To address this challenge, there is a need for more preclinical and clinical studies on EV-based therapies. This can include the development of standardized protocols for EV-based therapies, as well as the collection of long-term safety and efficacy data. Demonstration of the safety and efficacy of EV-based therapies will be essential for their approval and widespread clinical use.

8.3.5 Regulatory Challenges

There are several regulatory challenges that need to be addressed before EV-based therapies can be approved for clinical use. Currently, there are no specific regulatory guidelines for EV-based therapies, and they are typically regulated as biological products or drug delivery systems. However, the complex nature of EVs makes their regulatory approval more challenging.

One regulatory challenge is the lack of a standardized definition of EVs, which makes it difficult to develop consistent regulatory guidelines. The ISEVs has developed guidelines for EV characterization and nomenclature, but further efforts are needed to standardize EV classification and characterization. In addition, there are regulatory challenges related to the storage and transportation of EVs. EVs are sensitive to changes in temperature, and can be degraded during storage and transportation. This makes it challenging to maintain product quality during the manufacturing and distribution process. Finally, there are intellectual property issues related to EV-based therapies. Patents can be difficult to obtain for naturally occurring products like EVs, and there may be disputes over ownership and licensing agreements for EV isolation and characterization methods.

To address these regulatory challenges, there is a need for increased collaboration among researchers, regulatory agencies, and industry partners. The development of standardized EV characterization methods, manufacturing processes, and safety assessments will be essential to facilitate regulatory approval of EV-based therapies. In addition, there is a need for increased education and awareness of EV-based therapies among regulatory agencies and healthcare providers. This will help to ensure that these therapies are regulated appropriately, and that patients have access to safe and effective treatments.

8.3.6 Heterogeneity of EV Populations

Another challenge in EV research is the heterogeneity of EV populations. EVs can vary in size, shape, and composition depending on their source and method of isolation. This heterogeneity can complicate the interpretation of EV research, as it can be difficult to determine the functional significance of different EV populations.

To address this challenge, there is a need for a better understanding of the heterogeneity of EV populations. This can include the development of methods to isolate specific subpopulations of EVs, such as exosomes or microvesicles, as well as the characterization of the functional properties of different EV populations. Understanding the heterogeneity of EV populations will help to identify the most relevant EV populations for specific biomedical applications.

8.3.7 Understanding the Role of EVs in Disease Progression and Development

Finally, understanding the role of EVs in disease progression and development is another important challenge in EV research. While there is growing evidence for the involvement of EVs in various disease processes, there is still much that is unknown about their specific mechanisms of action.

To address this challenge, there is a need for more research on the role of EVs in disease progression and development. This can include the development of disease-specific EV biomarkers as well as the investigation of the mechanisms by which EVs contribute to disease pathogenesis. Understanding the specific cargo and signaling pathways of EVs in different diseases can provide important insights into disease mechanisms and help identify new therapeutic targets.

One approach to studying the role of EVs in disease is to perform large-scale EV profiling studies using techniques such as proteomics, lipidomics, and RNA sequencing. These studies can provide a comprehensive analysis of the EV cargo and identify potential biomarkers or therapeutic targets.

Another approach is to investigate the specific interactions between EVs and target cells in disease models. This can include studying the uptake and effects of EVs on recipient cells *in vitro* and *in vivo*, as well as assessing the efficacy of EV-based therapies in preclinical models.

In conclusion, while EV-based therapies hold great promise for the treatment of a wide range of diseases, there are several challenges that need to be addressed before they can be approved for clinical use. The development of standardized characterization methods, manufacturing processes, and safety assessments, as well as increased education and awareness, will be essential to overcome these challenges and realize the full potential of EV-based therapies.

Index

Biomedical Applications of Extracellular Vesicles, First Edition.
Edited by Zhenhua Li, Xing-Jie Liang, and Ke Cheng.

d

e